FOOD CHEMISTRY

主　编　冯凤琴
副主编　陆柏益　张　辉　刘松柏

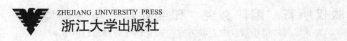

图书在版编目（CIP）数据

食品化学：Food Chemistry：英文/冯凤琴主编.
—杭州：浙江大学出版社，2013.9(2023.8 重印)
ISBN 978-7-308-12122-4

Ⅰ.①食… Ⅱ.①冯… Ⅲ.①食品化学－英文
Ⅳ.①TS201.2

中国版本图书馆 CIP 数据核字（2013）第 196150 号

FOOD CHEMISTRY

主　编　冯凤琴

责任编辑	阮海潮　严少洁
封面设计	姚燕鸣
出版发行	浙江大学出版社
	（杭州市天目山路148号　邮政编码310007）
	（网址：http://www.zjupress.com）
排　　版	杭州青翊图文设计有限公司
印　　刷	广东虎彩云印刷有限公司绍兴分公司
开　　本	787mm×1092mm　1/16
印　　张	15.5
字　　数	516 千
版 印 次	2013 年 9 月第 1 版　2023 年 8 月第 5 次印刷
书　　号	ISBN 978-7-308-12122-4
定　　价	42.00 元

版权所有　翻印必究　印装差错　负责调换

浙江大学出版社市场运营中心联系方式：0571-88925591；http://zjdxcbs.tmall.com

前　言

食品化学是食品科学相关专业的重要基础课程之一。食品化学的内容包括：食品及其原材料的组成成分，主要成分的结构及性质，这些成分在食品加工和保藏中的物理、化学及生物学变化，以及食品成分的结构、性质和变化对食品质量、安全及加工性能的影响等。食品科学与工程及相关专业的学生，必须掌握食品化学的基本知识和研究方法，才能在食品相关领域较好地从事教学、科研、生产及管理工作。

在全球化的背景下，越来越多的高等院校开始重视国际化人才培养，国内一些有条件的高等院校食品学院相继开展了食品化学的双语或英语教学，而目前国内还没有针对本科生食品化学教学的双语或英语教学教材。鉴于此，我们尝试编写了本书。本书内容在深度和广度上基本与食品科学与工程专业（本科）食品化学教学大纲相适应，同时也尽可能将食品化学领域的最新进展融入教材相关章节中。因此，本书除了可作为本科教材外，还可以供食品及相关学科的教师和研究生以及从事食品研究、开发和生产的技术人员参考。

参与本书编写的有浙江大学食品科学与营养系冯凤琴（第一、五、八章）、陆柏益（第六、七、九章）、刘松柏（第二、三章）和张辉（第四、十章），全书由冯凤琴主审。

由于编者水平有限，书中难免有错误和不妥之处，敬请读者批评指正。

CONTENTS

Chapter 1　Introduction ·· 1

1.1　**Definition** ·· 1
　1.1.1　Food and Food Science ··· 1
　1.1.2　Food Chemistry ··· 1
　1.1.3　Relationship between Food Chemistry and Other Disciplines ············ 2
1.2　**Content and Development of Food Chemistry** ·································· 2
1.3　**Approach to the Study of Food Chemistry** ····································· 3
　1.3.1　Quality and Safety Attributes ··· 3
　1.3.2　Chemical and Biochemical Reactions ···································· 4
　1.3.3　Effect of Reactions on the Quality and Safety of Food ········ 5
　1.3.4　Solve Problems by Analyzing and Controlling the Important Variables ··· 6
　Glossary ·· 7

Chapter 2　Water ·· 9

2.1　**Introduction** ·· 9
2.2　**Physical Properties and Structure of Water** ······································ 9
　2.2.1　Physical Properties ··· 9
　2.2.2　Structure of the Water Molecule ·· 11
　2.2.3　Water Intermolecular Interaction ··· 12
　2.2.4　Architecture of Water ·· 13
2.3　**Quantitative Description of Water in Foods** ···································· 15
　2.3.1　Water Content ··· 16
　2.3.2　Water Activity ··· 17
　2.3.3　Molecular Mobility ··· 18
2.4　**Water Activity and Food Properties** ··· 19
　2.4.1　Freezing ·· 19
　2.4.2　Combined Methods Approach to Food Stability ··················· 21
　Glossary ·· 23

FOOD CHEMISTRY

Chapter 3　Carbohydrate 26

3.1　Introduction 26
   3.1.1　Definition 26
   3.1.2　Classification 26
   3.1.3　Function and Distribution 27

3.2　Physical and Chemical Properties 28
   3.2.1　Physical Properties 28
   3.2.2　Chemical Properties 30

3.3　Common Sugars 33
   3.3.1　Monosaccharides 33
   3.3.2　Oligosaccharides 34
   3.3.3　Polysaccharide 37
   3.3.4　Starch 40
   3.3.5　Cellulose 43
   3.3.6　Pectin 44
   Glossary 45

Chapter 4　Protein 49

4.1　Introduction 49
   4.1.1　Definition 49
   4.1.2　Classification 50

4.2　Composition 51
   4.2.1　Structure 51
   4.2.2　Amino Acids 52

4.3　Properties 54
   4.3.1　Denaturation 54
   4.3.2　Gelation 59
   4.3.3　Emulsifying Properties 61
   4.3.4　Foaming Properties 64

4.4　Food Proteins 65
   4.4.1　Animal Proteins 65
   4.4.2　Plant Proteins 71

4.5　Peptides 72
   4.5.1　Properties 72
   4.5.2　Bioactive Peptides in Food 72
   Glossary 73

Chapter 5 Lipids ········ 78

5.1 Introduction ········ 78
    5.1.1 Definition and Roles ········ 78
    5.1.2 Classification of Lipids ········ 79
    5.1.3 Components of Triacyglycerols ········ 79
    5.1.4 Edible Oils ········ 81

5.2 Physical Properties ········ 83
    5.2.1 Melting Point and Polymorphism ········ 83
    5.2.2 Emulsions and Emulsifiers ········ 87

5.3 Chemical Reactivity of Fats and Oils ········ 89
    5.3.1 Oxidation ········ 89
    5.3.2 Other Reaction ········ 95

5.4 Functional Lipids ········ 98
    5.4.1 Functional Fatty Acids ········ 98
    5.4.2 Phytosterol ········ 99
    5.4.3 Phospholipids ········ 100

5.5 Fat Replacers and Mimetics ········ 100
    5.5.1 Fat Substitutes ········ 101
    5.5.2 Fat Mimetics ········ 101
    Glossary ········ 102

Chapter 6 Vitamins ········ 106

6.1 Introduction ········ 106

6.2 Fat-Soluble Vitamins ········ 107
    6.2.1 Vitamin A ········ 107
    6.2.2 Vitamin D ········ 109
    6.2.3 Vitamin E ········ 110
    6.2.4 Vitamin K ········ 111

6.3 Water-Soluble Vitamins ········ 115
    6.3.1 Ascorbic Acid ········ 115
    6.3.2 Thiamine (Vitamin B_1) ········ 119
    6.3.3 Riboflavin (Vitamin B_2) ········ 122
    6.3.4 Nicotinamide (Niacin) ········ 123
    6.3.5 Pyridoxine (Pyridoxal, Vitamin B_6) ········ 123
    6.3.6 Folic Acid ········ 124
    6.3.7 Biotin ········ 125
    6.3.8 Pantothenic Acid ········ 126
    6.3.9 Cyanocobalamin (Vitamin B_{12}) ········ 127

Glossary 128

Chapter 7　Enzymes 130

7.1　Introduction 130
7.2　General Remarks 130
 7.2.1　Catalysis 130
 7.2.2　Specificity 131
 7.2.3　Structures 131
 7.2.4　Nomenclature 132
 7.2.5　Activity Units 133
 7.2.6　Determination of Enzyme Activity 133
 7.2.7　Reasons for Catalytic Activity 134
 7.2.8　Kinetics of Enzyme-Catalyzed Reactions 135
7.3　Enzymes in Food 141
 7.3.1　Amylases 141
 7.3.2　Pectic Enzyme 145
 7.3.3　Cellulases and Hemicellulases 147
 7.3.4　Proteases 147
 7.3.5　Lipases 150
7.4　Enzyme Utilization in the Food Industry 151
 7.4.1　Starch 151
 7.4.2　Baking 151
 7.4.3　Brewing 152
 7.4.4　Cheese (Flavour) 152
 7.4.5　Coffee & Tea 152
 7.4.6　Dietetics 153
 7.4.7　Egg 153
 7.4.8　Juice & Wine 153
 Glossary 155

Chapter 8　Colorants 158

8.1　Introduction 158
8.2　Natural Pigments 159
 8.2.1　Chlorophyll 159
 8.2.2　Heme Compounds 164
 8.2.3　Carotenoids 168
 8.2.4　Phenolic Compound 172
8.3　Synthetic Colorants 181
 Glossary 184

Chapter 9 Flavors 188

9.1 Introduction 188
9.2 Taste and Nonspecific Saporous Sensations 188
    9.2.1 Taste Substances: Sweet, Bitter, Sour, and Salty 189
    9.2.2 Flavor Enhancers 190
    9.2.3 Astringency 191
    9.2.4 Pungency 192
    9.2.5 Cooling 192
9.3 Vegetable, Fruit, and Spice Flavors 193
    9.3.1 Sulfur-Containing Volatiles in Allium *sp.* 193
    9.3.2 Sulfur-Containing Volatiles in the Cruciferae 194
    9.3.3 Unique Sulfur Compound in Shiitake Mushrooms 196
    9.3.4 Methoxy Alkyl Pyrazine Volatiles in Vegetables 196
    9.3.5 Enzymically Derived Volatiles from Fatty Acids 197
    9.3.6 Volatiles from Branched-Chain Amino Acids 199
    9.3.7 Flavors Derived from the Shikimic Acid Pathway 199
    9.3.8 Volatile Terpenoids in Flavors 200
9.4 Flavor Volatiles in Muscle Foods and Milk 202
    9.4.1 Species-Related Flavors of Meats and Milk from Ruminants 202
    9.4.2 Species-Related Flavors of Meats from Nonruminants 204
    9.4.3 Volatiles in Fish and Seafood Flavors 205
9.5 Development of Process or Reaction Flavor Volatiles 206
    9.5.1 Thermally Induced Process Flavors 207
    9.5.2 Volatiles Derived from Oxidative Cleavage of Carotenoids 211
9.6 Future Directions of Flavor Chemistry and Technology 212
    Glossary 213

Chapter 10 Food Additive 214

10.1 Introduction 214
    10.1.1 Definition 214
    10.1.2 Classification 214
10.2 Acids 215
10.3 Bases 217
10.4 Buffer Systems and Salts 218
    10.4.1 Buffers and pH Control in Foods 218
    10.4.2 Salts in Processed Dairy Foods 219
10.5 Antimicrobial Agents 220
    10.5.1 Acid Antimicrobial 220

10.5.2　Ester Antimicrobial ······ 223
10.5.3　Inorganic Antimicrobial ······ 224
10.5.4　Biological Antimicrobial ······ 225
10.6　Sweeteners ······ 226
10.6.1　Saccharin ······ 226
10.6.2　Aspartame ······ 226
10.6.3　Acesulfame K ······ 228
10.6.4　Cyclamate ······ 228
10.7　Emulsifier ······ 229
10.7.1　Emulsifier Action ······ 230
10.7.2　Synthetic Emulsifiers ······ 233
10.8　Antioxidants ······ 235
Glossary ······ 235
References ······ 237

Chapter 1 Introduction

1.1 Definition

1.1.1 Food and Food Science

Food is concerned throughout the world, but the aspects of concerns about food differ with location. In underdeveloped regions of the world, the attainment of adequate amounts and kinds of basic nutrients remains an ever-present problem. In developed regions of the world, food is available in abundance, much of it is processed, and the use of chemical additives is common. In these fortunate localities, concerns about food relate mainly to cost, quality, variety, convenience, and the effects of processing and added chemicals on wholesomeness and nutritive value. All of these concerns fall within the realm of food science—a science that deals with the physical, chemical, and biological properties of foods as they relate to stability, cost, quality, processing, safety, nutritive value, wholesomeness, and convenience. Food science is an interdisciplinary subject involving primarily microbiology, chemistry, biology, and engineering.

1.1.2 Food Chemistry

Food chemistry, a major aspect of food science, is the application of chemistry principles to the food system. Food chemistry deals with the compositions and properties of foods as well as the chemical changes which undergoes during handling, processing, and storage. Chemistry is found at all levels in the food system; therefore, the scope of food chemistry is broad. This can be showed graphically by drawing lines from "chemistry" to each component of the food system. It is showed that food chemistry extends to agronomy, harvesting, extraction, processing and/or refining, packaging, storage, distribution, and retail (Fig. 1.1). We need to "use" food chemistry to study the behaviors of food ingredients within the whole food system.

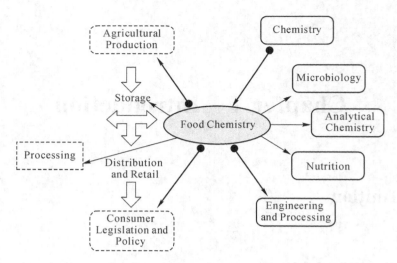

Fig. 1.1 Relating food chemistry with the food system

1.1.3 Relationship between Food Chemistry and Other Disciplines

Food chemistry is intimately related to chemistry, biochemistry, physiological chemistry, botany, zoology, and molecular biology. Food chemists rely heavily on knowledge of the aforementioned sciences to effectively study and control biological substances as sources of human food. Knowledge of the innate properties of biological substances and mastery of the means of manipulating them are common interests of both food chemists and biological scientists. The primary interests of biological scientists include reproduction, growth, and changes that biological substances undergo under environmental conditions that are compatible or marginally compatible with life. On the contrary, food chemists are concerned primarily with biological substances that are dead or dying (postharvest physiology of plants and postmortem physiology of muscle) and changes they undergo when exposed to a very wide range of environmental conditions. In addition, food chemists are concerned with the chemical properties of disrupted food tissues (flour, fruit and vegetable juices, isolated and modified constituents, and manufactured foods), single-cell sources of food (eggs and microorganisms), and one major biological fluid, milk. In summary, food chemists have much in common with biological scientists, but they also have interests that are distinctly different and are of the utmost importance to humankind.

1.2 Content and Development of Food Chemistry

Food is complicated as for its components. There is an emphasis an food chemistry, including macro-constituents (water, carbohydrates, lipids, and proteins), micro-constituents (for example, flavors, vitamins, minerals, pigments), and their interactions.

Food chemistry emerged as a discipline after World War II, and its initial mission was to ensure the supply of food which is nutritious, safe, and affordable. This mission is shared by the other disciplines of food sciences including food microbiology, food processing, food engineering, and food laws. With the fast development and advancement of economy and human society in the past decades, there was a new emphasis on food processing as to make the consumption of food more convenient and the chemical transformations encountered during processing to improve the quality of processed foods. Therefore, advances were made in understanding the maillard reaction, which leads to colored and flavored compounds during processing. The effect of moisture content on food deterioration was also examined leading to a prolonged shelf-life of stored foods. The behaviors of food hydrocolloids, including starch, pectin, bacterial polysaccharides, and protein gelatin, were elucidated. More and more food additives such as sweeteners, antimicrobials, antioxidants, flavorings were developed and applied in food to improve the sensory attributes and to ensure the safety of food. Developments in food analysis enabled the detection of pesticide residues and other toxicants in food.

1.3 Approach to the Study of Food Chemistry

It is desirable to establish an analytical approach in food chemistry for food formulation, processing, and storage stability, so that the derived facts from the study of one food or model system can help us to resolve the problems appeared in other food products. There are four components to this approach: (a) to determine what are the properties that are important characteristics of safe, high quality food, namely quality attributes, (b) to determine what are the chemical and biochemical reactions that have important influences on loss of quality and/or wholesomeness of foods, (c) to integrate the aforementioned two points to understand how the key chemical and biochemical reactions influence the quality and safety of foods, and (d) to apply this understanding to various situations encountered during formulation, processing and storage of foods.

1.3.1 Quality and Safety Attributes

It is important to reiterate that safety is the first requisite of any food. In a broad sense, safety means a food must be free of any harmful chemical or microbial contaminant at the time of its consumption. In practical operation, this definition takes on a more applied form. For example, in the canning industry, "commercial" sterility as applied to low acid food means the absence of viable spores of *Clostridium botulinum*. This in turn can be translated into a specific set of heating conditions for a specific product in a specific package. Given these heating requirements, one can select specific time-temperature conditions that will optimize retention of quality attributes. Similarly, for peanut butter, operational safety can be regarded primarily as the absence of aflatoxins-a strong carcinogenic substance produced by certain species of molds. Measures taken to prevent growth of the mold may or may not interfere with the retention of some other quality attributes, nevertheless, steps producing a safe product must be employed.

The quality attributes of food including texture, flavor, color, nutritive value and safety and some alterations they may undergo during handling, processing or storage are listed in Table 1.1. The changes that may occur, with the exception of those involving nutritive value and safety, are readily evident to consumers.

Table 1.1 Classification of alterations that may occur on food attributes during handling, processing or storage

Attribute	Alteration
Texture	Loss of solubility
	Loss of water-holding capacity
	Toughening
	Softening
Flavor	Development of:
	Rancidity (hydrolytic or oxidative)
	Cooked or caramel flavors
	Other off-flavors
	Desirable flavors
Color	Darkening
	Bleaching
	Development of other off-colors
	Development of desirable colors (e.g. browning of baked goods)
Nutritive value	Loss, degradation or altered bioavailability of proteins, lipids, vitamins, minerals
Safety	Generation of toxic substances
	Development of substances that are protective to health
	Inactivation of toxic substances

1.3.2 Chemical and Biochemical Reactions

Many reactions can alter food quality or safety. Some important classes of these reactions are listed in Table 1.2. Each type of the reaction may involve different reactants or substrates depending on the specific food and the particular conditions for handling, processing, or storage. They are treated as reaction types because the general nature of the substrates or reactants is similar for all foods. For example, non-enzymatic browning involves reaction of carbonyl compounds, which can arise from existing reducing sugars or from diverse reactions, such as oxidation of ascorbic acid, hydrolysis of starch, or oxidation of lipids. Oxidation may involve lipids, proteins, vitamins, and pigments, and oxidation of lipids may involve triacylglycerols in one food or phospholipids in another. Discussions of these reactions in detail will be carried out in subsequent chapters of this book.

Table 1.2 Some chemical and biochemical reactions that can lead to alteration of food quality or safety

Types of reaction	Examples
Non-enzymic browning	Baked goods
Enzymic browning	Cut fruits
Oxidation	Lipids (off-flavors), vitamin degradtion, pigment decoloration, proteins (loss of nutritive value)
Hydrolysis	Lipids, proteins, vitamins, carbohydrates, pigments
Metal intereactions	Complexation (anthocyanins), loss of Mg from chlorophyll, catalysis of oxidation
Lipid isomerization	Cis→trans, nonconjugated→conjugated
Lipid cyclization	Monocyclic fatty acids
Lipid polymerization	Foaming during deep fat frying
Protein denaturation	Egg white coagulation, enzyme inactivation
Protein cross-linking	Loss of nutritive value during alkali processing
Polysaccharide synthesis	In plant postharvest
Glycolytic changes	Animal tissue postmortem, plant tissue postharvest

1.3.3 Effect of Reactions on the Quality and Safety of Food

The reactions listed in Table 1.3 cause the alterations listed in Table 1.1. Integration of information contained in both tables can lead to an understanding of the causes of food deterioration. Deterioration of food is usually caused by a series of primary events, i.e., physical changes or chemical reactions, followed by secondary events, which, in turn, become evident alteration of quality attributes (Table 1.1) ultimately. Examples of the sequences of this type are shown in Table 1.3. Note particularly that a given quality attribute may be altered as a result of several different primary events.

The sequence in Table 1.3 may be applied in two directions. Operating from left to right, one can design a particular primary, associated secondary events, and get the desirable quality attributes. Alternatively, one can determine the probable cause(s) of an observed quality change (column 3, Table 1.3) by analysing all primarily events that could be involved and then isolating, by appropriate chemical tests, the key primary event. The utility of such constructed sequences is that one can approach the problems of food alteration in an analytical manner.

FOOD CHEMISTRY

Table 1.3 Cause and effect relationship pertaining to food alteration during handling, processing or storage

Primary causative event	Secondary event	Attribute influenced (See Table 1.1)
Hydrolysis of lipids	Free fatty acids react with protein	Texture, flavor, nutritive value
Hydrolysis of polysaccharides	Sugars react with proteins	Texture, flavor, color, nutritive value
Oxidation of lipids	Oxidation products react with many other constituents	Texture, flavor, color, nutritive value; toxic substances can be generated
Bruising of fruit	Cells break, enzymes are released, oxygen accessible	Texture, flavor, color, nutritive value
Heating of green vegetables	Cell walls and membranes lose integrity, acids are released, enzymes become inactive	Texure, flavor, color, nutritive value
Heating of muscle tissue	Proteins denature and aggregate, enzymes become inactive	Texture, flavor, color, nutritive value
Cis → trans con → versions in lipids	Enhanced rate of polymerization during deep fat frying	Excessive foaming during deep fat frying; diminished bioavailability of lipids

1.3.4 Solve Problems by Analyzing and Controlling the Important Variables

Having a description of the attributes of high-quality, safe foods, the significant chemical reactions involved in the deterioration of food and the relationship between the two, we can now begin to consider how to apply these information to the situations encountered during the processing and storage of food, which involves analyzing and controlling the important variables including temperature, time, pH, water activity (a_w), light, composition of the product, and composition of the atmosphere and so on.

Temperature is perhaps the most important among these variables because of its broad influences on all types of chemical reactions. Another important factor is time. During storage of a food product, one is interested in time which is respected to the integral of chemical and/or microbiological changes that occur during a specified storage period, and these changes combine together to determine a specified storage life for the product. During processing, one is often interested in the time it takes to reach an ideal sterilization effect or in how long it takes for a reaction to proceed to a specified extent.

Another variable, pH, influences the rates of many chemical and enzymatic reactions. Extreme pH values are usually required for severe inhibition of microbial growth or enzymatic processes, and these conditions may accelerate the acid or base catalyzed reactions. In contrast, even a relatively small pH change may cause profound changes in the quality of some foods, for example, muscle.

Another important compositional determinant of reaction rates in foods is water activity (a_w). Numerous investigators have shown a_w strongly influences the rate of enzyme-catalyzed reactions, lipid oxidation, non-enzymatic browning, sucrose hydrolysis, chlorophyll degradation, anthocyanin degradation, and others.

Recently, it has become apparent that the glass transition temperature (T_g) of food and the corresponding water content (W_g) of the food at T_g are causatively related to rates of diffusion-limited events in food. Thus T_g and W_g have relevance to the physical properties of frozen and dried foods, to conditions appropriate for freeze drying, to physical changes involving crystallization, recrystallization, gelalinization, and starch retrogradation, and to those chemical reactions that are diffusion-limited.

For some products, exposure to light can cause detrimentation, and it is then appropriate to package the products in light-proof material or to control the intensity and wavelengths of light, if possible.

Composition of the product is important since this determines the reactants available for chemical transformation. In some processed foods, the composition can be controlled by adding approved chemicals (food additives), such as acidulants, chelating agents, flavors, or antioxidants, or by removing undesirable reactants, for example, removing glucose from dehydrated egg albumen.

Composition of the atmosphere is important mainly with respect to relative humidity and oxygen content, although ethylene and CO_2 are also important during storage of living plant foods. The detrimental consequences of a small amount of residual oxygen sometimes become apparent during storage of food product.

Food chemists must integrate the information about quality attributes of foods, deteriorative reactions to which foods are susceptible, and the factors governing kinds and rates of these deteriorative reactions, in order to solve problems related to food formulation, processing, and storage stability.

Glossary

aflatoxins	黄曲霉毒素
albumen	清蛋白
anthocyanin	花青素
antioxidant	抗氧化剂
ascorbic acid	抗坏血酸,维生素 C
botany	植物学
carbohydrate	糖类
carcinogenic	致癌的

FOOD CHEMISTRY

chelating agent	螯合剂
chlorophyll	叶绿素
clostridium botulinum	肉毒杆菌
commercial sterility	商业无菌
degradation	降解
dehydrated	脱水的
denaturation	变性
enzyme-catalyzed reaction	酶促反应
ethylene	乙烯
food deterioration	食物败坏
food ingredient	食品成分
gelatinization	凝胶化
glucose	葡萄糖
glycolytic	糖分解
humidity	湿度
hydrocolloid	凝胶,亲水胶体
hydrolytic/hydrolysis	水解的/水解
inhibition	抑制
isomerization	异构化
light-proof	不透光的
lipids	脂类
maillard reaction	美拉德反应
moisture content	水分含量
molecular biology	分子生物学
non-enzymitic browning	非酶褐变
pectin	果胶
pesticide residue	农药残留
phospholipid	磷脂
pigments	色素
polymerization	聚合
polysaccharide	多糖
postharvest physiology	采后生理学
postmortem physiology	尸检生理学
protein gelatin	蛋白质明胶
rancidity	酸败,食品腐败
recrystallization	重结晶
spore	孢子
starch	淀粉
sucrose	蔗糖
susceptible	灵敏的,易受感染的
synthesis	合成
toxicants	有毒物质
triacylglycerol	三酰基甘油

Chapter 2 Water

2.1 Introduction

Water is ubiquitous on the earth and is the medium of life. Water affects properties of food fundamentally in many aspects. It interferes with the texture of food as a lubricant and plasticizer contributing moistness and disturbing solute-solute interactions. It is powerful and chemically inert solvent for flavors, colorants, nutrients, salts, and the substances essential for life such as proteins (enzymes), nucleic acids, sugars determining the conformations and facilitating the dynamic behavior of bio-macromolecules. It also directly participates in many processes by supplying protons or hydroxyls. As a consequence, its state of presence greatly influences the growth of microbes, enzyme activity and food properties accordingly.

The unique role of water in living processes and the environment attributes to its remarkable properties. Water does not dissolve non-polar substances such as lipids, which is indispensable for the origin of life. Just through hydrophobic interactions, lipids form membranes and the cellular nature of life is established. Water's thermal properties especially contribute to its environment fitness. Its high heat capacity makes it possible to buffer climate and thermal change. The anomalous expansion of water as it cools to 0 ℃ forms an insulating ice cover on the surface protecting the liquid water underneath. In the following sections, water's physical and chemical properties will be reviewed and their implications in food properties will be examined subsequently.

2.2 Physical Properties and Structure of Water

2.2.1 Physical Properties

Water normally exists in nature in three common states: liquid, solid, and gaseous. It boils at 100 ℃ (212 ℉, 373 K) and freezes at 0 ℃ (32 ℉, 273 K) under the standard pressure. However, if keeping well undisturbed pure liquid water can be super-cooled to approximately $-42\ ℃$ (231 K) without freezing. The boiling point of water is substantially dependent on its external pressure. For instance, water deep in the ocean can remain liquid with the temperatures reaching hundreds of degrees. The fourth state of water, a supercritical fluid, is much less common. In this state, water achieves a specific critical temperature (647 K) and a specific critical pressure (22.064 MPa). Under these conditions, the two phases (liquid and gas) merge to one homogeneous fluid phase, so that the supercritical water has unusual properties of both gas and liquid.

There is appreciable energy flow as water transforms between different states. Liquid water owns the second highest heat capacity (4.18 J/g) of any known substance (after ammonia), ten times greater than that of iron. It also has the second highest (after ammonia) heat of fusion at $0\ ℃$ (335 J/g). It has exceptionally high heat of vaporization (2.24 kJ/g). These unusual thermodynamic properties suggesting strong intermolecular interactions confer the capacity to buffer climate fluctuation and to maintain constant body temperature. Before the advent of refrigerators, ice still is common used to retard food spoilage. The density of water is dependent on its temperature, approximately one gram per cubic centimeter. The precise relation is complicated, neither linear nor monotonic. When cooled from room temperature, liquid water becomes increasingly dense and reaches its maximum density at $3.98\ ℃$. As cooled further, water becomes less dense. This unusual negative thermal expansion results from strong, orientation-dependent, intermolecular interactions. These properties have important outcomes for lives on the earth. Water at $4\ ℃$ always accumulates at the bottom of lakes and ice floats on the surface. Because ice is poor heat conductor, it preserves aquatic life.

Water has extraordinarily high dielectric constant ($\varepsilon = 80.18$, $20\ ℃$) rendering it exceptionally good solvent for ionization. Various ions in solution enrich chemical properties and functionalities in life and food systems. As a generally accepted concept, all the substances can be divided into two types, hydrophilic and hydrophobic, according to solubility in water. Substances that dissolve in water are hydrophilic (water-loving), while those that don't dissolve in water are hydrophobic (water-fearing). The solubility of a substance is determined by the ability of the substance molecules to compensate the strong intermolecular interaction between water molecules. If a substance cannot match the strong interaction in water, the molecules will be expelled out of the network of water molecules and do not dissolve. Generally ionic and polar molecules such as acids, alcohols, proteins, and salts can dissolve in water to certain extent and non-polar molecules don't dissolve forming membranes.

Due to strong intermolecular interaction water molecules tend to stay close to each other, called cohesion. The strong cohesion between water molecules endows water's surface tension (72.8 mN/m at room temperature) which is the highest in all common liquids (except mercury). Because of water molecule's polar nature, water molecules would interact strongly with hydrophilic molecules like glass (adhesion). An interplay of the forces of cohesion and adhesion accounts for considerable part of water's role in food stuff such as capillary action, a phenomenon that water rises in a narrow tube against the force of gravity, and hydration of protein or membrane surfaces.

It is the most marvelous thing that all the unusual properties occur on a single substance, water. The key to understanding the remarkable properties of water lies in the strong intermolecular interaction due to its unraveled capability to form hydrogen bonds.

2.2.2 Structure of the Water Molecule

The strong intermolecular interactions between water molecules exhibit uncommon structures for water. To understand these features, the nature of a single water molecule will be considered first and then small groups of molecules. The bond angle of a water molecule is 104.5°, which is close to the perfect tetrahedral angle of 109° 28′. The internuclear distance for O—H is 0.96 Å and the van der Waals radii for oxygen and hydrogen are 1.40 and 1.2 Å, respectively. Thus, a water molecule has V-like shape which is originated from the nature of sp^3-hybridization of orbitals of the oxygen's outermost shell. Two hydrogen atoms form two covalent σ-bonds with the oxygen via two sp^3 orbitals, each of which has a dissociation energy of 460 kJ/mol. The localized molecular orbitals are symmetrically oriented about the original orbital axes, thus adopting an approximate tetrahedral structure. A schematic orbital model of a water molecule is shown in Fig. 2.1.

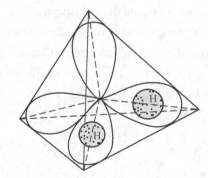

Fig. 2.1 Schematic model of a single HOH molecule: sp^3 hybridization

The V-like geometry of a water molecule and the polarity of the O—H bond result in an unsymmetrical charge distribution in the whole molecule. Consequently pure water owns a dipole moment of 1.84 D in vapor-state. Polarity produces intermolecular attractive forces between water molecules. However, the large dipole moment of water only accounts for a fraction of its unusually large intermolecular forces. This unusual intermolecular force can only be understood on the basis of extensive three-dimensional hydrogen bonding between water molecules. As a matter of fact, water constituents are far more complicated than the picture presented so far. Except for ordinary HOH molecules pure water also contains many other constituents in minute amounts. There are 18 isotopic variants of HOH, ^{17}O, ^{18}O, ^{2}H (deuterium) and ^{3}H (tritium) present together with the common isotopes ^{16}O and ^{1}H. Some ionic entities also exist in water such as hydrogen ions (existing as H_3O^+), hydroxyl ions, and their isotopic variants. Therefore more than 33 chemical variants of HOH are present in water, but those variants occur in only trace amounts.

2.2.3 Water Intermolecular Interaction

Water's ability to engage in multiple hydrogen bonding three-dimensionally quite adequately explains its large intermolecular attractive forces. A hydrogen bond forms between a hydrogen bond donor (a hydrogen atom covalently bonded to an electronegative atom, typically O or N) and a hydrogen bond acceptor (an electronegative atom, typically O or N). Hydrogen bonds, at a strength of 12 to 30 kJ/mol, are much weaker than covalent bonds (typically 335 kJ/mol), but stronger than normal intermolecular forces (van der Waals). In fact hydrogen bonding forms straight bonding between the donor atom and the acceptor atom and is highly directional. For this reason, hydrogen bonding is more specific than van der Waals interaction and requires the presence of complementary hydrogen donor and acceptor groups.

The electron pairs in the two O—H bonds of the water molecule are drawn away from the two hydrogens by the highly electronegative oxygen, leaving two partially positive charged hydrogens and one partially negative charged oxygen. The orientation of hydrogen bonding of water molecules takes a tetrahedral configuration. Two axes of an imaginary tetrahedron assume the two hydrogen-oxygen bonding orbitals (hydrogen bond donor sites). The remaining two axes of the imaginary tetrahedron adopt oxygen's two lone-pair orbitals (hydrogen bond acceptor sites). By virtue of these imaginary four lines, each water molecule is able to form four hydrogen bonds. The schematic tetrahedral arrangement is depicted in Fig. 2.2.

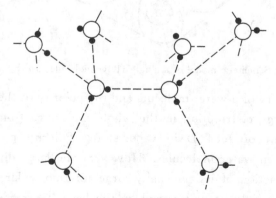

Fig. 2.2 Hydrogen bonding of water molecules in a tetrahedral configuration. Open circles are oxygen atoms and closed circles are hydrogen atoms. Hydrogen bonds are represented by dashed lines

Each water molecule has an equal number of hydrogen bonding donor and acceptor sites, and therefore permits formation of three dimensional hydrogen bonded networks. This nature is responsible for water molecules' unusually large intermolecular attractive forces, when compared to other small molecules strongly engaging in hydrogen bonding such as NH_3 and HF. Ammonia has three hydrogens (donor sites) and one receptor site, and hydrogen fluoride has one hydrogen (donor site) and three receptor sites. Without equal numbers of donor and receptor sites they can form only two dimensional hydrogen bonded networks. Inclusion of isotopic variants and hydronium and hydroxyl ions greatly complicates the picture of water's hydrogen bonding network. The positively charged hydronium ions exhibit greater hydrogen-bond donating potential than un-ionized water molecules. The negatively charged hydroxyl ions exhibit greater hydrogen-bond acceptor potential than un-ionized water.

Water also hydrogen bonds with other polar molecules. In food, water molecules attract sugars, polysaccharides, and proteins via hydrogen bonding. Although the most important intermolecular force is hydrogen bonding, the water molecule is still a permanent dipole strongly attracting with other permanent or induced dipoles through dipole-dipole interaction. The interaction energy between permanent dipoles is stronger than that between a permanent dipole and an induced dipole. The former falls off as $1/r^3$, whereas the latter falls off as $1/r^4$. Ionized water molecules interact attractively with charged molecules like salts and proteins via electrostatic interaction. Because hydrophobic molecules cannot form hydrogen-bonds, their presence would disrupt water's hydrogen bonding network, and then a reshuffle must happen.

2.2.4 Architecture of Water

Liquid water has the open, hydrogen-bonded, and tetrahedral network. Obviously this structure is not sufficiently established to generate long-range rigidity, but far more organized than that in the vapor state. Thus the behavior of a water molecule is amply influenced by its neighbors. If water is a nonstructured liquid and assumes the close packing mode, its density would be 70% higher than that it actually is. Partial retention of the hydrogen-bonded arrangement of ice easily accounts for water's low density. The high heat of fusion of ice is believed to break only about 15% of the hydrogen bonds exist in ice.

The degree of hydrogen bonding in water is certainly temperature dependent. In ice each water molecule has a coordination number (number of nearest neighbors) of 4.0, with a nearest intermolecular distance of 2.76 Å (Fig. 2.3). In the process of melting some hydrogen bonds are broken with an increase of the distance between nearest neighbors (decreased density) and an increase of the coordination number (increased density). As the temperature rises, the distance between nearest neighbors increases from 2.76 Å to 2.9 Å at 1.5 °C, then to 3.05 Å at 83 °C. Simultaneously, the coordination number increases from 4.0 to 4.4 at 1.50 ℃, and then to 4.9 at 83 ℃. At the early stage of ice-to-water transformation (0 to 3.98 ℃) the factor of the increase of the coordination number predominates to result in a net increase in density. At 3.98 ℃ water reaches its maximum density. Further warming above that point the factor of the increase of the distance between nearest neighbors predominates and the density declines.

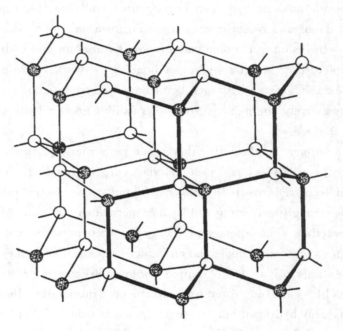

Fig. 2.3　Three dimensional hydrogen bonding network of water

　　It is not self-evident that water has relatively low viscosity considering the presence of strong hydrogen bonding interaction. This paradox is readily reconcilable with the highly dynamic hydrogen bonding relationships between neighboring molecules within the time frame of nano- to picoseconds, thereby facilitating mobility and fluidity.

　　To elucidate the structure of pure water is extremely complicated. Many theories have been proposed, but all are incomplete and far from perfect. Three general types of models will be introduced: mixture, interstitial, and continuum. Mixture models employ the concept that intermolecular hydrogen bonds are momentarily (about 10^{-11} sec), and concentrated in dynamically equilibrated clusters of water molecules. Interstitial models involve the idea that water retains a clathrate-type (ice like) structure and the interstitial spaces of the clathrates are filled with individual water molecules(Fig. 2.3). Continuum models apply the notion that hydrogen bonds are distributed uniformly in ice, and as ice melts the existing bonds simply become distorted rather than broken, allowing a picture of a dynamic continuous network of water molecules. The common structural feature in all three models is the hydrogen-bonded association of liquid water in ephemeral, distorted tetrahedron. All models also permit individual water molecules to frequently exchange for a new hydrogen bond and alter their bonding arrangements, while maintaining a constant degree of hydrogen bonding and structure for the entire system.

2.3 Quantitative Description of Water in Foods

The fact that the existing state of water has profound effect on food's properties, such as stability and palatability, which has long been recognized. To prevent spoilage, minimize quality loss, ensure food safety and lengthen shelf-life are important tasks for food scientists. To achieve this goal, water management is invoked as a unifying concept. Before the discussion, a few basic concepts must be clarified. The terms "water binding" and "hydration" are often employed to describe a tendency for water to associate with hydrophilic substances, i.e., proteins, sugars and cellular materials. A number of factors including nonaqueous constituent, salt composition, pH, and temperature determine the degree and tenacity of water binding or hydration. "Water holding capacity" is frequently used to describe the ability of a matrix of molecules, including gels of pectin and starch, and cells of tissues, to physically entrap large quantities of water in a way that inhibits exudation.

On the basis of the degree of water binding water can be divided into bulk water (the same as pure water), associated water (loosely bound) and bound water (tightly bound). Because bound water is not an easily identifiable entity, this term is controversial, and poorly understood. Although increasing numbers of scientists suggest terminating its use, the term "bound water" must be discussed owing to its common presence in the literature.

There are numerous definitions proposed for "bound water" as follows.

1. Bound water is the equilibrium water content of a sample at some appropriate temperature and low relative humidity.

2. Bound water is that what not contribute significantly to permittivity at high frequencies and therefore has its rotational mobility restricted by the substance with which it is associated.

3. Bound water is what does not freeze at some arbitrary low temperature (usually -40 ℃ or lower).

4. Bound water is what is unavailable as a solvent for additional solutes.

5. Bound water is what produces line broadening in experiments involving proton nuclear magnetic resonance.

6. Bound water is what moves with a macromolecule in experiments involving sedimentation rates, viscosity, or diffusion.

7. Bound water is that which exists in the vicinity of solutes and other nonaqueous substances and has properties differing significantly from those of "bulk" water in the

same system.

When a given sample is examined, these definitions usually produce different values. In a typical food of high water content, this type of water comprises only a minute part of the total water, approximately. It is useful to view bound water as the first layer of water molecules in the vicinity of solutes and other nonaqueous constituents. Bound water should not be regarded as immobilized but with diminished mobility.

To quantitatively describe water in food, several complimentary approaches have been established: (ⅰ)Water content, which is a traditional method and simple in concept and measurement; (ⅱ) Water activity (a_w), which accounts for effects of solutes and differences between foods; (ⅲ)Molecular mobility, which treats food as polymer matrixs, and water as key plasticizer.

2.3.1 Water Content

Water content is usually defined as the weight percentage of water in food. Many foods have their own characteristic of water content (Table 2.1). Water content profoundly influences textures, appearances, and flavors of food and their susceptibility to spoilage. It has long been recognized that the water content of food positively correlate with their perishability. To decrease the water content and perishability accordingly, concentration and dehydration processes are carried out primarily.

Since most kinds of fresh food own large water content, effective forms of preservation are required to achieve long-term storage. Reduction of water content greatly changes the native properties of food and biological activities of food regardless of what approaches applied, either conventional or freezing dehydration. In addition, returning water to its original status after dehydration is even more challenging. All attempts including rehydration and thawing are never more than partially successful.

Table 2.1 Water contents of various foods

Food	Water content (%)
Meat	
Pork, raw, composite of lean cuts	53-60
Beef, raw, retail cuts	50-70
Chicken, all classes, raw meat without skin	74
Fish, muscle proteins	65-81
Fruit	
Berries, cherries, pears	80-85
Apples, peaches, oranges, grapefruit	90-95
Rhubarb, strawberries, tomatocs	90-95
Vegetables	
Avocado, bananas, pears (green)	75-80
Beets, broccoli, carrots, potatoes	85-90
Asparagus, bean (green), cabbage, cauliflower, lettuce	90-95

2.3.2 Water Activity

In the practice of water management, it has been found that the term of water content alone is not sufficient to reflect all aspects of water properties. In many cases food with the same water content differs significantly in perishability. This phenomenon originates from a fundamental principle that the availability of water in food determines food perishability. Once water associates with nonaqueous constituents, water partially loses its capability to support degradative activities including growth of microorganisms and hydrolytic chemical reactions, namely losing its availability. The stronger water associates with nonaqueous constituents, the more availability it loses. The term "water activity" (a_w) was developed to describe the intensity with which water associates with various nonaqueous constituents.

Based on a_w various food properties like food stability and food safety can be predicted far more reliably than from water content. Even so, a_w is not a completely reliable indicator. Despite this imperfection, a_w is sufficiently good to be a useful indicator of product stability and microbial safety. It correlates well enough with rates of microbial growth and many degradative reactions. Actually a_w is specified in some U.S. federal regulations involving good manufacturing practices(aMp) for food.

As pioneered by Scott, water activity was studied by applying the general notion of substance "activity" which was rigorously derived from the laws of equilibrium thermodynamics by G. N. Lewis. It is sufficient here to state that

$$a_w = f/f_0 \tag{2-1}$$

where f is the fugacity of the solvent in solution (fugacity is the escaping tendency of a solvent from solution) and f_0 is the fugacity of the pure solvent. Under low pressure (i.e., ambient), the difference between f/f_0 and P/P_0 (where P is the vapor pressure of the solvent and P_0 is the vapor pressure of the pure solvent) is less than 1%, so defining a_w in terms of P/P_0 is readily justifiable. Therefore,

$$a_w = P/P_0 \tag{2-2}$$

Solution ideality and the existence of thermodynamic equilibrium are prerequisites for establishment of this equality. Unfortunately both assumptions are generally violated in food. As a result, Equation 2-2 must be taken as an approximation and the proper expression is

$$a_w \approx P/P_0 \tag{2-3}$$

Since P/P_0 sometimes does not equal a_w, it is more reasonable to directly employ the term P/P_0 rather than a_w. The term P/P_0 is named "relative vapor pressure" (RVP), and the two terms are used interchangeably.

To study the relationship of water content and RVP, a moisture sorption isotherm (MSI) is developed that is a plot of water content of a food versus P/P_0 at constant temperature. Plenty useful information can be derived from MSIs, including (ⅰ) information on the ease or difficulty of water removal essential for concentration and dehydration processes, (ⅱ) information for formulating food mixtures to minimize moisture transfer among the ingredients, (ⅲ) information on the moisture barrier properties required in packaging, (ⅳ) information for determining what moisture content to inhibit growth of microorganisms of interest, and (ⅴ) information for forecasting the chemical and physical stability of food as a function of water content.

A schematic MSI for a high moisture food is shown in Fig. 2.4, the full range of water content from normal to dry included in this plot. At very low moisture content (water content), water is strongly bounded with solutes and is unavailable, and a_w is low ($a_w<0.2$). As moisture content increases, some water becomes free from solutes and a_w increases. It has to be pointed out that a_w is temperature dependent and a_w increases as temperature increases.

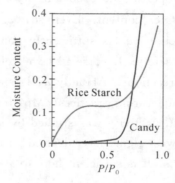

Fig. 2.4 MSI for a high moisture food (van den Berg and Bruin, 1981)

2.3.3 Molecular Mobility

Although the RVP approach has served as reliable indicator for the food industry, in the cases where diffusion becomes a limiting factor, molecular mobility (Mm, translational or rotational motion) must be taken into account complimentarily to effectively predict and control food stability and processability. Most physical properties of foods are diffusion-limiting, and most chemical properties are controlled by chemical reactivity. However, there do exist some diffusion limiting chemical properties. Complicated approaches have been developed to predict that chemical reactions are limited by diffusion or chemical reactivity. Due to the importance of this approach careful studies should be continued. The RVP approach deals with "availability" of water in foods, i.e., to which degree water can function as a solvent, while the Mm approach deals with microviscosity and diffusibility of chemicals in foods. These approaches complement each other to predict and control food properties.

2.4 Water Activity and Food Properties

Water activity has become an important and reliable indicator for food quality and spoilage. The two factors, water content and water activity, together influence significantly the progress of chemical and microbiological spoilage reactions in foods. Typical dried or freeze-dried foods with great storage stability usually possess water contents in the range of about 5% to 15%. Foods with intermediate moisture, like cakes, have water contents in the range of about 20% to 40%. The dried foods correspond to the lower part of the moisture sorption isotherms. This includes water in the monolayer and multi-layer category (bound water). Intermediate-moisture foods usually have a_w above 0.5 including the capillary water. Reduction of water activity can be achieved by drying or by adding hydrophilic substances, i.e., sugar to jams, or salt to pickled preserves. As a_w below 0.90 bacterial growth barely happens. Even though some osmophile yeast strains can grow at low a_w (down to 0.65), molds and yeasts usually can be inhibited when a_w drops to between 0.88 and 0.80.

2.4.1 Freezing

Freezing is regarded as the best method for long-term preservation of most kinds of foods. One fact has to be clarified that benefits of the preservation technique of freezing derive primarily from low temperature not from ice formation. Furthermore, the formation of ice in cellular foods and food gels actually has two adverse consequences: (a) nonaqueous constituents are concentrated in the unfrozen phase and (b) water to ice transformation causes 9% expansion in volume.

During freezing, water from aqueous solutions, cellular suspensions, or tissues is transformed into ice crystals of high purity. Thus, nonaqueous constituents are concentrated in less quantity of unfrozen water. The degree of concentration is determined mainly by the final temperature, and to a lesser degree by rate of cooling, agitation, and formation of eutectics (crystallization of solutes—uncommon).

T_m^l is the melting point curve; T_E is the eutectic point; T_m^s is the solubility curve; T_g is the glass transition curve; T'_g is the solute-specific glass transition temperature of a maximally freeze concentrated solution.

The freeze-concentration effect greatly changes concentration-related properties of the unfrozen phase including pH, titratable acidity, ionic strength, viscosity, surface and interfacial tension, oxidation-reduction potential, and colligative properties. Furthermore, water-water and water-solute interactions may be dramatically altered, solutes may crystallize, supersaturated oxygen and carbon dioxide may be expelled from solution, and

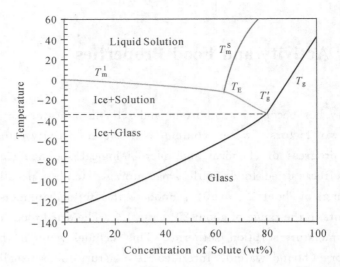

Fig. 2.5 A simplified binary temperature-composition phase diagram(Adapted from Roos and Karel, 1999)

macromolecules will be forced to pack closer together. There are two opposing factors on concentration-related reaction rate during freezing: (a) lowered temperature, decreasing reaction rates, and (b) freeze-concentration, increasing reaction rates.

The discussion of Mm and stability of frozen food or food with reduced moisture content can be significantly facilitated by consideration of phase diagrams. A simplified binary temperature-composition phase diagram is shown in Fig. 2.5. The two lines, glass transition curve (T_g) and the line extending from T_E to T_g, in phase diagram represent metastable conditions. Most samples existing in a state of nonequilibrium are located above the glass transition curve and not on any line.

When these diagrams are applied, constant pressure and negligible time dependency of the metastable states are assumed. The phase diagram in Fig. 2.5 provides a unified qualitative description of simple systems although each will always have its own characteristic phase diagram that differs quantitatively. Most foods are so complex that they cannot be accurately or easily represented by a state diagram. Although accurate determination of a glass transition curve of complex foods, either dry or frozen, is difficult, estimating of T_g could be done with accuracy sufficient for commercial use on the basis of the Mm approaching for food stability.

Establishment of the equilibrium curves (T_m^l and T_m^S; Fig. 2.5) for complex foods is also challenging. For the major equilibrium curve of dry or semidry foods, the T_m^S curve usually cannot be accurately represented by a single line. It is commonly used that deducing properties of the complex food from the phase diagram for water and a food solute of dominating importance to the properties of the complex food. For instance, the properties and behavior of cookies during baking and storage can be credibly predicted by a phase diagram for sucrose-water. If a food solute of dominating importance cannot be found in dry or semidry complex foods, determining T_m^S curves will be much more difficult and still remains a problem to be resolved.

Because the melting point curve (T_m^l, the major equilibrium curve of importance) is often known or easily determined, it is relatively simple to prepare a phase diagram for a complex frozen food with accuracy sufficient for commercial purposes.

2.4.2 Combined Methods Approach to Food Stability

At this point, it is hopefully established that RVP and Mm are powerful tools for food stability but that either or even both are totally sufficient. The drawbacks of these approaches have to be noted: (a) the RVP approach alone to control microbial growth is often inadequate, (b) the RVP approach is not a totally reliable predictor of chemical stability because it is only a function of a single parameter, and (c) the Mm approach is also not a totally reliable predictor of chemical stability because it is also only a function of a single parameter. Other factors not accommodated by either of the two approaches may have important influences on food stability and safety. Thus, the "combined methods approach" was developed to control microbial growth in foods in complicated real food production.

To obtain conditions required to inhibit the growth of microorganisms in nonsterile foods Professor L. Leistner and others developed the combined methods approach (originally called "hurdle approach"). This approach manipulates the sequence of various growth controlling parameters to completely suppress the growth where each parameter is a "hurdle" to microbial growth. This approach is best illustrated by the examples in Fig. 2.6. The dashed, undulating line indicates the progress of a microorganism attempting to overcome the inhibitory hurdles, and the growth occurs only after all hurdles have been overcome. Size of the hurdle represents relative inhibitory effectiveness. In real life microorganisms obviously would confront all hurdles simultaneously rather than in sequence as is shown, and some of the factors would function synergistically.

Six hurdles are present in Example 1 and growth is sufficiently inhibited because the microorganisms are unable to overcome all hurdles. A typical microbial population is present in Example 2 and RVP and preservatives are the most potent of the various hurdles with different inhibitory effectiveness, achieving satisfactory control of microbial growth. In Example 3, the same product and the same hurdles are present with a small starting population of microorganisms that result from good sanitary practices. In this example, fewer hurdles suffice. In Example 4, the same product and the same hurdles are present with a large starting population of microorganisms that result from poor sanitary practices. In this example, the hurdles are insufficient to realize satisfactory control of microbial growth. Example 5 provides the same hurdles and population of organisms as in Example 2 but with the sample rich in nutrients. Owing to the nutrients, the hurdles that were adequate in Example 2 are inadequate in this instance. In Example 6, the microorganisms are exposed to a substerilizing treatment before storage with the product

FOOD CHEMISTRY

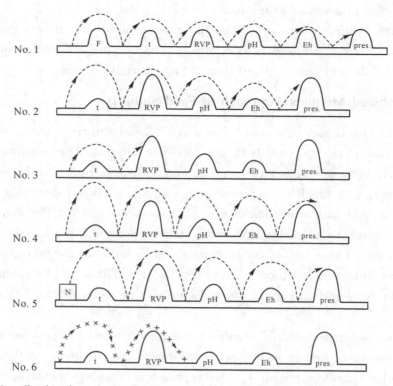

Fig. 2.6 Combined methods approach to food stability. The combined methods approach to controlling growth of microorganisms in nonsterilized food. F is heating, t is chilling, RVP is relative vapor pressure, pH is acidification, Eh is redox potential, pres. is chemical preservative, and N is nutrients. Closed circle—hurdles in foods; dashed lines—population of microorganism (Adapted from Leistner, 1995)

and hurdles unchanged. Because of the substerilizing treatment, the surviving organisms are damaged and are less able to overcome the hurdles so that fewer hurdles suffice.

In conclusion, the following has to be borne in mind. Although RVP and Mm are powerful tools for predicting and controlling the properties and stability of food, there exit many occasions where either or even both are totally sufficient and other factors, including chemical properties of the solute, pH, and oxidation-reduction potential, must also be considered.

Glossary

agitation	搅动
ammonia	氨
amply	充足地
anomalous expansion	反常膨胀
arbitrary	任意的；任意角度
asparagus	芦笋；龙须菜；天冬
buffer	缓冲
capillary action	毛细管作用
cauliflower	花椰菜，菜花
clathrate	包合物，笼形物
cohesion	凝聚，结合；内聚力
colligative properties	依数性质
compensate	补偿，抵消
configuration	配置；结构；外形
continuum	连续统一体
coordination number	配位数
covalence	共价；共价键
covalent	共价的，共有原子价的
covalent bonds	共价键
depict	描述
dielectric constant	介电常数，电容率
diffusion	扩散，传播，漫射
dipole moment	偶极矩
dissociation energy	离解能，分解能
dynamic	动态的，动力的
electron pair	电子对
electronegative	带负电的，负电性的
electrostatic interaction	静电作用，静电相互作用
elucidate	阐明，说明
ephemeral	短暂的
equation	方程式，等式，反应式

equilibrate	平衡,使平衡,与平衡
equilibrium	均衡;平静;保持平衡的能力
eutectic crystallization	共晶结晶
exudation	渗出;渗出物,分泌,分泌物
facilitating	促进
fluctuation	起伏,波动
fugacity	逸度,逸散能
fundamentally	根本地,从根本上,基础地
heat of fusion	熔解热,熔化热
hydration	水合作用
hydrogen bonding	氢键,氢键结合
hydrolytic	水解的,水解作用的
hydronium	水合氢离子
hydrophobic interaction	疏水作用
hydroxyls	氢氧根
implication	含义,暗示
inert	惰性的
insulating	绝缘的,隔热的;使绝缘
intermolecular	分子间的;作用于分子间的
isotopic	同位素的;同位旋的
localized	局部的;地区的;小范围的;定位
low viscosity	低黏度
lubricant	润滑剂;润滑油;润滑的
macromolecule	高分子;大分子
marvelous	了不起的;非凡的;令人惊异的
matrix	基质
membrane	细胞膜
mercury	水银;水银柱
microviscosity	微黏度
mobility	移动性;机动性;[电子]迁移率
monolayer	单层;单层的
monotonic	单调的;无变化的
negligible	微不足道的,可以忽略的
nonaqueous	非水的
nucleic acids	核酸;核苷酸
orbital	轨道的

osmophile	嗜高渗菌
oxidation-reduction potential	氧化还原电位;氧化还原势
palatability	适口性;风味
parameter	参数;系数;参量
pectin	果胶;胶质
perishability	易腐性
permittivity	介电常数,电容率
pickled	腌制的;盐渍的
plasticizer	塑化剂;可塑剂
polymer	聚合物
polysaccharides	多糖
predominate	占主导地位;占优势
proton	质子
retention	保留;扣留,滞留
rigidity	硬度,刚性
rotational	转动的;回转的;轮流的
schematic	原理图;图解视图;概要
sedimentation rate	沉降速度
solvent	溶剂;有溶解力的
sorption isotherm	等温吸附线
supercritical fluid	超临界流体
susceptibility	敏感性
tetrahedral	四面体的;有四面的
tetrahedron	四面体
thermodynamic property	热力学性质
thermodynamics	热力学
three-dimensional	三维的;立体的
titratable acidity	滴定酸度;可滴定酸度
ubiquitous	普遍存在的;无所不在的
undulating	波状的;波浪起伏的;使波动
unravel	解决;散开
unsymmetrical	非对称的;不匀称的;不调和

Chapter 3 Carbohydrate

3.1 Introduction

3.1.1 Definition

Carbohydrates from "hydrate of carbon" are the most abundant biomolecules of organism, and sometimes referred to as saccharides. They have the empirical formula $C_m(H_2O)_n$ (where m could be different from n), and signify molecules containing carbon atoms, hydrogen and oxygen atoms, with a hydrogen : oxygen atom ratio of 2 : 1, which is the same ratio as they occur in water. With further research, some saccharides do not comply with the above formula, such as rhamnose and ribodesose, and some saccharides contain other compositions, such as nitrogen, phosphorus and sulfur, but people still use the name-carbohydrate. The names of carbohydrates often end in the suffix -ose. Chemically, carbohydrates are organic compounds which are more accurate to be defined as aldehydes or ketones of many hydroxyl groups (Fig. 3.1).

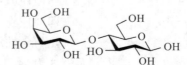

Fig. 3.1 The structure of lactose

3.1.2 Classification

Carbohydrates can be classified depending on different properties, which are outlined in Table 3.1.

Generally, carbohydrates are classified as three groups: monosaccharides, oligosaccharides and polysaccharides. Monosaccharides are the simplest carbohydrates because they cannot be hydrolyzed to smaller carbohydrates. Monosaccharides are commonly referred simply as sugars. However, as will be explained later, table sugar (sucrose) is not a monosaccharide. The monosaccharides can be further classified according to the number of carbon atoms contained, as shown in Table 3.1.

Table 3.1 Classification of carbohydrate

No.	Property	Classification			
1	Complexity	Simple carbohydrates monosaccharides		Complex carbohydrates oligosaccharides & polysaccharides	
2	Number of carbon atoms	C_4 sugars Tetrose	C_5 sugars Pentose	C_6 sugars Hexose	C_7 sugars Heptose
3	C=O function	Aldose [sugars having an aldehydes function or an acetal equivalent] Ketose [sugars having a ketone function or an ketal equivalent]			
4	Reactivity	Reducing [sugars can be oxidized by Tollen's reagent (or Benedict's or Fehling reagents)] Non-reducing [sugars cannot be oxidized by Tollens' or other reagents]			

An oligosaccharide typically contains 2 to 10 monosaccharide units joined by glycosidic bonds. According to the number of constituent monosaccharides, oligosaccharide can be subdivided into disaccharide, trisaccharide, tetrasaccharide and so on, in which the disaccharide is the most important, i. e., sucrose, maltose. The polysaccharide contains greater than ten monosaccharide units. When a polysaccharide is composed of two or more different monosaccharide units, it is a heteroglycan. Conversely, when it contains same monosaccharide units, it is called homopolysaccharide. Monosaccharide derivates, such as amino sugar and uronic acid, also are the compositions of polysaccharides.

3.1.3 Function and Distribution

Carbohydrate occurs abundantly in the organisms, which comprise more than 90% of the dry mass of plants. Food carbohydrates include sugars, starches, gums, fiber, and so on. We use the sugar percentages of total fruit weight as an example, the details are shown in Table 3.2.

Table 3.2 Different sugar percentages of total fruit weight

Fruit \ Sugar percentages(%)	Glucose	Fructose	Sucrose	Total
Apple	1.17	6.04	3.78	10.99
Grape	6.68	7.84	2.25	16.95
Peach	0.91	1.18	6.92	9.01
Pear	0.95	6.77	1.61	9.30
Cherry	6.49	7.38	0.22	14.09
Strawberry	2.09	2.40	1.03	5.22

The main functions of carbohydrates are:

1. They are human energy sources, providing 80% calories in human diet and energy for metabolism pathways and cycles.

2. They are the basis for synthesis of other compounds.
3. They are structural components for cells and tissues.
4. They are signals for cell recognition and identification of biological molecules.

3.2 Physical and Chemical Properties

3.2.1 Physical Properties

Of many physical properties of sugar, three most important properties including optical activity, sweetness and water-holding capacity will be discussed in detail in this section. Optical activity is the ability of a chiral molecule to rotate the plane of plane-polarised light. Carbohydrates are optically active compounds. In general, an optically active molecule will rotate the plane of polarization through an angle that is proportional to the thickness of the sample and to the concentration of the molecule. The sugar concentration in a solution can be measured conveniently by using the optical activity of sugar and other asymmetric molecules. Optical rotation is an important indicator for carbohydrate. Some sugar optical rotations are listed in Table 3.3.

Table 3.3 Optical rotations of some sugar*

Sugar	Optical rotation	Sugar	Optical rotation
D-glucose	+52.2	D-mannose	+14.2
D-fructose	−92.4	D-arabinose	−105.0
D-galactose	+80.2	D-xylose	+18.8
L-arabinose	+104.5		

* Concentration of solution is 1 g/1 mL; the thickness is 0.1 m; temperature is 20 ℃

In order to introduce optical activity clearly, we use monosaccharide with the simplest structure as an example. Glyceraldehyde, which is possessing only one chiral carbon atom, is the simplest aldose with chemical formula $C_3H_6O_3$. In the nature, there are two kinds of glyceraldehyde isomers—"right-handed form" of glyceraldehyde and "left-handed form" of glyceraldehyde, denoted as D-glyceraldehyde and L-glyceraldehyde; however, most of glyceraldehydes are in the D configuration. Structures of the two isomers are shown in Fig. 3.2.

The structures of isomers are written according to a convention developed by Emil Fischer who is the founder of many monosaccharides. By convention, the aldehyde or ketone is set at the top. The hydroxyl group on the second-to-last carbon atom is drawn on the right side of skeleton structure in case of the D-isomer and on the left for the L-isomer. The two isomers of glyceraldehyde written by the Fischer project are shown in

```
         CHO                    CHO
    H ──┼── OH            HO ──┼── H
        CH₂OH                  CH₂OH
    D-glyceraldehyde       L-glyceraldehyde
```

Fig. 3.2 Structures of D-isomer and L-isomer

Fig. 3. 2. L-sugars rarely exist in nature, but nevertheless have important biochemical roles. Two L-sugars, L-arabinose and L-galactose, are found in foods as units in polysaccharides.

In solution, a single molecule can interchange between straight and ring form. The alcohol at one end of a monosaccharide can attack the carbonyl group at the other end to form a cyclic compound. The —OH groups can act as nuceophile, which can change the size of ring. As a result, there are two kinds of heterocyclic ring, pyranose with six carbons and furanose with five carbons, as shown in Fig. 3. 3. A pyranose or a furanose also has two possible structures, which are called the α- and the β-anomers. The open isomer D-glucose gives rise to four distinct cyclic isomers: α-D-glucopyranose, β-D-glucopyranose, α-D-glucofuranose, and β-D-glucofuranose; which are all chiral.

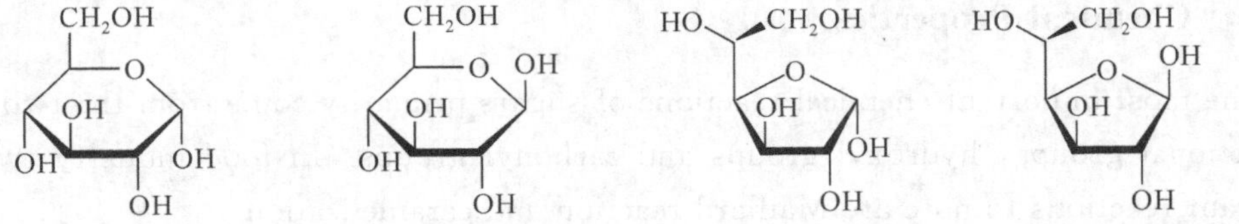

α-D-glucopyranose   β-D-glucopyranose   α-D-glucofuranose   β-D-glucofuranose

Fig. 3.3 The cyclic isomers of D-glucose

Sweetness is an important physical property of sugar. However, the strength of sweetness can not be quantitative determination by chemical or physical means, so sensory evaluation is used to compare sweetness. Sweetness of sucrose is usually used as base material, and sucrose in solution (10% or 15% sucrose solution, 20 ℃) has a sweetness perception rating of 1, and other substances are rated relative to this, so the value obtained is relative, as shown in Table 3. 4.

Table 3.4 Sweetness of various saccharides

Name	Sweetness
Sucrose	1 (reference)
α-D-glucose	0.7
Frutose	1.17-1.75
Lactose	0.16
Maltose	0.33-0.45

The strength of sweetness is determined by the molecular structure, molecular weight, molecular state of existence and external factors. All of monosaccharides, most of the disaccharides, and some of trisaccharides have sweetness, but polysaccharides is not sweet. When the molecular weight becomes larger, the solubility becomes weaker, and the sweetness diminishes. The conformations of sugar also affect its sweetness. For example, if the sweetness of α-D-glucose is 1.0, the value of β-D-glucose is 0.666. Crystalline glucose is in the α-D-glucose, but some will be changed to β-D-glucose in the solution, so the sweetness will be the maximum at the beginning of the dissolved glucose.

Water-holding capacity (WHC) is defined as the ability of a food to retain its water during processing and storage. It affects the yield and juiciness of the final product. Many of the physical properties of foods, including color, texture and firmness are significantly influenced by their water-holding capacity. Cooked tenderness and juiciness are partially dependent on WHC. The WHC of molecules present in foods contribute to overall WHC of foods. Together with proteins, sugars account for major part of WHC of foods. The highly hydroxylated structure of sugars endows their great power to bind with water molecules through hydrogen-bonding. Water-holding capacity of sugars changes with the specific structures, other constituents like salts, pH and temperature, etc.

3.2.2 Chemical Properties

The most important chemical reactions of sugars primarily come from the two classes of functional groups, hydroxyl groups and carbonyl groups. In food industry two most important reactions to note are Maillard reaction and caramelization.

3.2.2.1 Maillard Reaction

Maillard reaction, known as hydroxylamine reaction, has played an important role in browning of food under some conditions. This reaction is named after the French scientist Louis Camille Maillard, who studied the reactions of amino acids and carbohydrates in 1912. The Maillard reaction is a form of nonenzymatic browning similar to caramelization. Almost all food contain carbonyl (from aldehyde or ketone) and amidogen (from protein), so the Maillard reaction may occur in all kinds of food processing. As the Maillard reaction is related to aroma, taste and color of foods, it has been a major challenge in food industry.

The chemistry underlying the Maillard reaction is very complicated. As aldoses or ketoses are heated with amines in solution through frying, roasting, baking, or storage, a chain of reactions ensue with production of numerous compounds. Some of them are either desirable or undesirable flavors, aromas, and deep-colored polymeric materials, but both reactants only slowly disappear.

The reducing sugar reacts reversibly with the amine to afford a glycosylamine, as illustrated with D-glucose (Fig. 3.4). This compound undergoes a process called the Amadori rearrangement to provide a derivative of 1-amino-1-deoxy-D-fructose in this case.

This Amadori product will proceed, especially at pH 5 or lower, with continuous dehydration to yield intermediates with multiple carbonyls. Eventually a furan derivative, 5-hydroxymethyl-2-furaldehyde (HMF), is formed in the case of a hexose (Fig. 3.5). Under less acidic conditions (higher than pH 5), the furan derivatives like HMF polymerize promptly to a dark-colored, insoluble nitrogen-containing material.

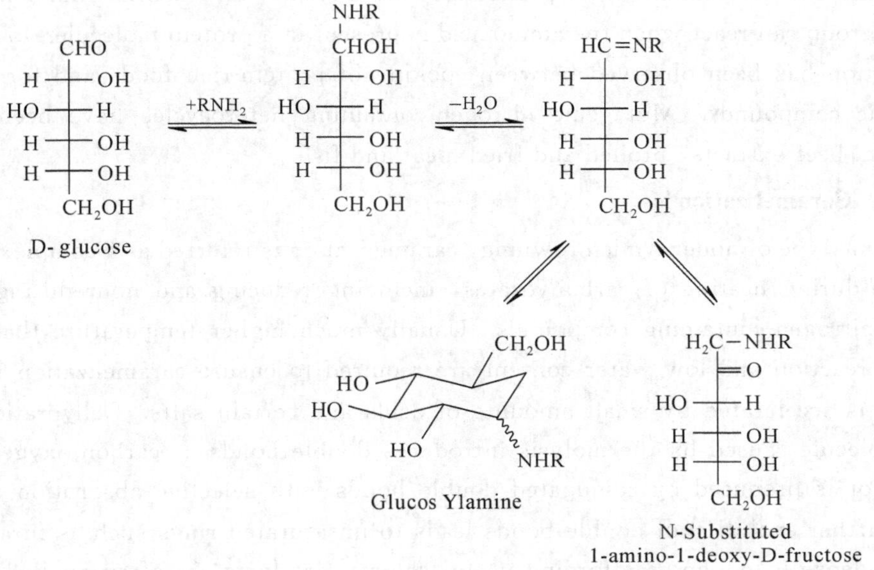

Fig. 3.4 Amadori rearrangement

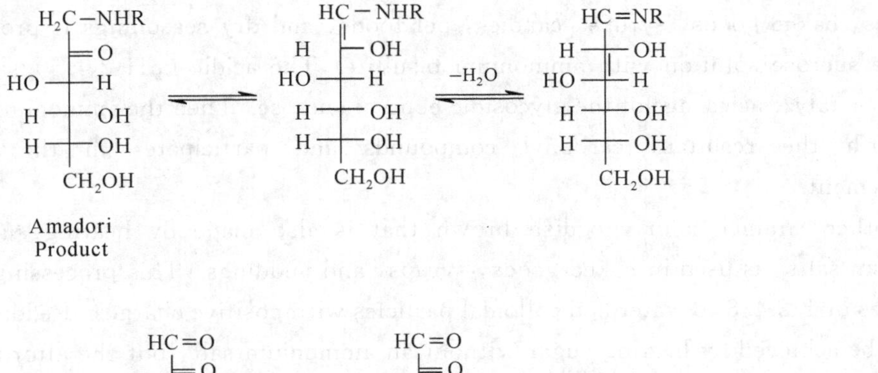

Fig. 3.5 Formation of 5-hydroxymethyl-2-furaldehyde

For nonenzymic browning employment of heat is generally required. Heating reducing sugars and amino acids, proteins, and/or other nitrogen-containing compounds together, such as in soy sauce and bread crusts, Maillard browning products are found including soluble and insoluble polymers. Maillard reaction products greatly contribute to the flavor of milk chocolate. The Maillard reaction of reducing sugars with milk proteins is also essential in the production of caramels, coffees, and fudges. Despite their advantages, Maillard reactions also have a negative side. Amino acids are destroyed during reaction of reducing sugars with amino acids, particularly for L-lysine, an essential amino acid whose ε-amino group can react when the amino acid is present in a protein molecule. In addition, a correlation has been observed between cooking of protein-rich foods and formation of mutagenic compounds. Mutagenic nitrogen-containing heterocycles have been isolated from from beef extracts, broiled and fried meat and fish.

3.2.2.2 Caramelization

As one type of nonenzymic browning, caramelization is referred as a complex group of reactions during heating of carbohydrates, including reducing and nonreducing sugars, without nitrogen-containing compounds. Usually much higher temperature than that in Maillard reaction and low water content are required to ensure caramelization to occur. Reaction is accelerated by small amounts of acids and certain salts. Dehydration of the sugar molecule caused by thermolysis introduces double bonds or carbon-oxygen bonds. Then color is produced by conjugated double bonds with selective absorption of visible light. Further evolution of double bonds leads to unsaturated rings such as furans which often condense to polymers affording useful colors. Catalysts are often applied to direct the reaction to specific types of caramel colors, solubilities, and acidities.

Brown caramel color which is widely used in Cola, soft drinks, other acidic beverages, baked goods, syrups, candies, pet foods, and dry seasonings is produced by heating a sucrose solution with ammonium bisulfite. The acidic (pH 2-4.5) salt in the solutions catalyzes cleavage of the glycosidic bond of sucrose. Then the ammonium ion will react with the resultant carbonyl compounds and participates in the Amadori rearrangement.

Another caramel color, reddish brown that is also made by heating sugar with ammonium salts, is used in baked goods, syrups, and puddings. This processing imparts pH values of 4.2-4.8 to water and colloidal particles with positive charges. Reddish brown can also be achieved by heating sugar without an ammonium salt, but the afforded syrup has a solution pH of 3-4 and consists of colloidal particles with slightly negative charges, which is used in beer and other alcoholic beverages. Large polymeric molecules with complex, variable, and unknown structures comprise caramel pigments. These polymers account for the colloidal particles whose rate of formation increases with elevated temperature and pH. Certain unique flavors and fragrances as well as the coloring materials can be yielded by pyrolytic reactions of specific sugars. For example, the flavor of bread results from pyrolytic products: maltol (3-hydroxy-2-methylpyran-4 one), isomaltol (3-hydraxy-2-acetylfuran), and 2h-4-hydroxy-5-methylfuran-3-one that enhances various flavors and sweeteners (Fig. 3.6).

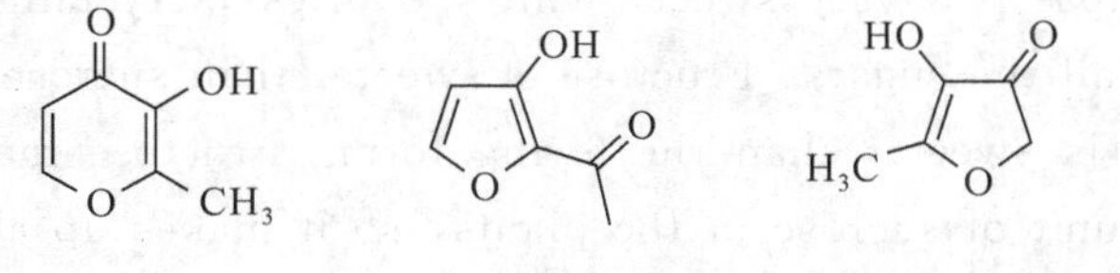

Maltol   Isomaltol   2h-4-hydroxy-5-
                                          methylfuran-3-one

Fig. 3.6 Pyrolytic flavor products of bread

3.3 Common Sugars

3.3.1 Monosaccharides

Monosaccharides are considered as simple sugars, because they cannot be broken down to simple carbohydrate molecules by hydrolysis. They contain a single aldehydes or ketones functional group, so they are subdivided into two classes-aldoses and ketoses-according to whether they are aldehydes or ketones (Table 3.5). In the nature, the simplest monosaccharides are glyceraldehydes and dihydroxyacetone, and the most important and common monosaccharides are glucose and fructose.

Table 3.5 Classification of monosaccharides

Number of carbon atoms	Kind of carbonyl group	
	Aldehyde	Ketone
3	Trioses	Triulose
4	Tetrose	Tetrulose
5	Pentose	Pentulose
6	Hexose	Hexulose
7	Heptose	Heptulose
8	Octose	Octulose
9	Nonose	Nonulose

Glucose is the most common hexose and a reducing sugar. It is a ubiquitous monosaccharide in cells and used as the primary energy source. Glucose is colorless and easily soluble in water, acetic acid, and several other solvents. Solutions of glucose rotate polarized light to the right, so called dextrose. The glucopyranose forms of glucose predominate in solution, and α- and β-D-glucopyranose are in equilibrium in aqueous solution. They melt at 146 ℃ (α) and 150 ℃ (β).

Pure, dry D-fructose is a very sweet, white, odorless, crystalline solid and is the most water-soluble of all the sugars. Fructose is sweeter than sucrose. However, the 5-ring form of fructose is sweeter than the 6-ring form. Fructose may be presented as monosaccharide or as unit of sucrose in the plants, so it makes up about 55% of high-fructose corn syrup and about 40% of honey. D-fructose has only three chiral carbon atoms, C-3, C-4, and C-5. Thus, there are only 2^3 or 8 D-ketohexoses.

3.3.2 Oligosaccharides

An oligosaccharide contains a small number of sugar units (typically two to ten), joined by glycosidic bonds.

A glycosidic bond is formed between the hemiacetal group (C1 of aldoses and C2 of ketoses) of a saccharide and the hydroxyl group of another saccharide, and removed a molecule of water. There are two kinds of glycosidic bonds: α- and β-glycosidic bond, which is decided by the position of —OH group on carbon-1. An α-glycosidic bond is formed when the —OH group is below the plane of glucose ring and a β-glycosidic bond is formed when it is above the plane. For example, maltose is formed of two glucose molecules linked by 1-4α-glycosidic bonds, and sucrose is formed by a glucose and a fructose linked by 1-2 β-glycosidic bond. Glycosidic bonds readily undergo acid catalyzed hydrolysis in the presence of aqueous acid and heat since they are part of acetal structures.

The simplest oligosaccharide contains two monosaccharide units named disaccharide. Trisaccharide includes three glycosyl units and the compound contains from 4 to 10 units, called tetra-, penta-, hexa-, octa-, nona-, and decasaccharides, and so on.

3.3.2.1 Sucrose

Sucrose is commonly known as table sugar and cane sugar, and is widely distributed in fruits, roots, stems, leaves, flowers and seeds of plants. In sugar industry, sugar cane and sugar beet are used as raw materials, in which sugar can account for 12%-20% of the plant's dry weight, to extract sucrose. Sucrose is the most important sweetener both in human lives and food industry, and about 150,000,000 metric tons are produced annually.

Sucrose is composed of one glucose unit and one fructose unit joined at C-1 and C-4 as a β-glycoside, as shown in Fig. 3.7. It is a nonreducing sugar because it has no anomeric hydroxyl groups.

Sucrose molecular formula is $C_{12}H_{22}O_{11}$, and the melting point is 186 ℃. Pure sucrose is a white, odorless, crystalline powder with a sweet taste. Sucrose can be soluble in water easily, and the solubility increases with temperature rise.

Sucrose, as a pure carbohydrate, has an energy content of 3.94 kilocalories per gram, providing most energy for human. It is also used extensively in fermentations, bakery products and pet food. In mammals' stomachs, sucrose is readily digested into D-glucose and D-fructose, which can be rapidly absorbed into the bloodstream in the small intestine.

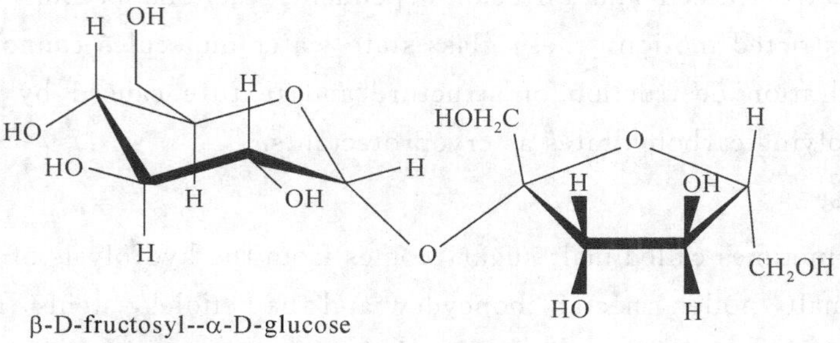

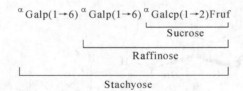

β-D-fructosyl--α-D-glucose
Sucrose

Fig. 3.7 The structure of sucrose

This step is acidic hydrolysis and performed by a glycoside hydrolase.

Raffinose (a trisaccharide which has a D-galactopyranosyl unit attached to sucrose) and stachyose (a tetrasaccharide which contains D-galactosyl unit) are found in sugar beet extract (Fig. 3.8). These oligosaccharides that are also present in beans, are nondigestible and are responsible for the flatulence due to eating beans.

$$^{\alpha}\text{Galp}(1\to 6)\,^{\alpha}\text{Galp}(1\to 6)\,^{\alpha}\text{Galcp}(1\to 2)\text{Fruf}$$

```
                                   |_____|
                                         Sucrose
                   |_____|
                                 Raffinose
 |_____|
                        Stachyose
```

Fig. 3.8 Sucrose, raffinose, and stachyose

Produced by white sugar crystals treated with molasses, commercial brown sugar is used as a coating of desired thickness. Its color ranges from light yellow to dark brown. Pulverized sucrose is usually applied in confection or powdered sugar that often contains 3% corn starch as an anticaking agent. To prepare fondant sugar utilized in icings and confections, very fine sucrose crystals are blended with a saturated solution of invert sugar, corn syrup, or maltodextrin.

Sucrose is not crystallized but shipped as a refined aqueous solution known as liquid sugar for many food product applications. Sucrose and most other carbohydrates with low molecular weight including monosaccharides, disaccharides, and some oligosaccharides, can form highly concentrated solutions of high osmolality due to their great hydrophilicity and solubility. As exemplified by pancake, waffle syrups and honey, such solutions need no preservatives themselves and can be used not only as sweeteners but also as preservatives and humectants.

A portion of water is nonfreezable in any carbohydrate solution. When the freezable water crystallizes, the freezing point of the remaining liquid phase decreases as the concentration of solute increases. Consequentially viscosity of the remaining solution increases. Eventually, the liquid phase solidifies as a glass where the movement of all molecules is greatly freezed and diffusion-dependent reactions become very restricted. Owing to the restricted motion, these glass-state water molecules cannot crystallize. In this way, dehydration destruction of structure and texture caused by freezing can be prevented by applying carbohydrates as cryoprotectants.

3.3.2.2 Maltose

Maltose, sometimes called malt sugar, comes from the hydrolysis of starch. Maltose presents in the malt, pollen, nectar, honeydew and the petioles, stems, roots of soybean plants. Maltose is generated during the dough fermentation and sweet potatoes baking, and used in brewing beer. Maltose is white powder or crystals, and the solubility in water is 1.080 g/mL (20 ℃). Maltose has a sweet taste, about half as sweet as glucose and about one-third as sweet as sucrose.

Maltose also has two monosaccharides units, two glucose units linked at C-1 and C-4 as a α-glycoside (Fig. 3.9). Since the compound has a potentially free aldehyde group which is the reducing end, it equilibrates with alpha and beta six-membered ring forms.

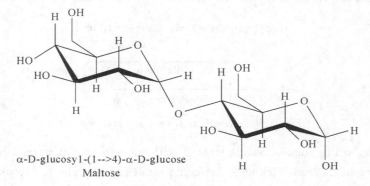

α-D-glucosyl-(1-->4)-α-D-glucose
Maltose

Fig. 3.9　Structure of maltose

3.3.2.3 Lactose

Lactose (Fig. 3.10) occurs in milk mainly as free disaccharide, but to certain extent as a component of higher oligosaccharides. It is the primary carbohydrate source for developing mammals. The concentration of lactose in milk ranges from 2.0% to 8.5% with the mammalian source. Human milks contain 7% lactose, and Cow and goat about 4.5%-4.8%. In humans, lactose provides 40% of the energy consumed during nursing. Preceding hydrolysis of lactose to the constituent monosaccharides must take place to use it for energy. Milk also contains 0.3%-0.6% of lactose-containing oligosaccharides which are important energy sources for growth of the predominant microorganism of the intestinal flora of breast-fed infants, a specific variant of Lactobacilluss bifidus.

Milk and other unfermented dairy products, i.e., ice cream supply lactose. Some of the lactose is converted into lactic acid during fermentation. Therefore, less lactose is contained in fermented dairy products including most yogurt and cheese. Lactose

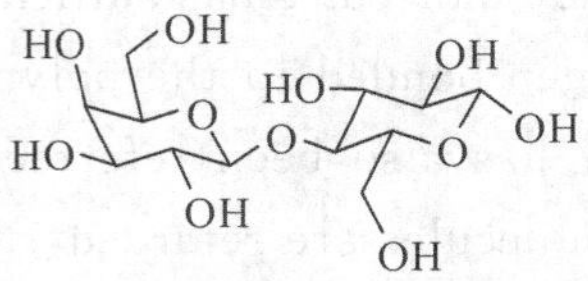

Fig. 3.10 Structure of lactose

stimulates intestinal adsorption and retention of calcium. The hydrolytic enzyme lactase is located in the brush border epithelial cells of the small intestine where lactose starts to be digested. The lactase (a β-galactosidase) is a membrane-bound enzyme that catalyzes the hydrolysis of lactose into its constituent monosaccharides, D-glucose and D-galactose which are rapidly absorbed and enter the blood stream.

3.3.3 Polysaccharide

Like the oligasaccharides, polysaccharides are composed of glycosyl units (more than 10 units) in linear or branched arrangements. The term, degree of polymerization (DP), is used to describe the number of monosaccharide units in a polysaccharide. Most polysaccharides have DPs in the range from 20 to 3,000, and only a few of them have a DP less than 100. Typically, cellulose has a DP of 7,000-15,000. It is estimated that more than 90% of the considerable carbohydrate mass in nature existing in the form of polysaccharides.

When a polysaccharide is composed of the same monosaccharide, all the glycosyl units are homogeneous in terms of monomer units and it is called a homoglycan. Examples of linear homoglycans are cellulose and starch amylose, and starch amylopectin is branched homoglycan. All three are constituted only with D-glucopyranosyl units. If the glycosyl units of a polysaccharide consist of two or more different monosaccharide types, it is a heteroglycan. A polysaccharide containing two different monosaccharide units is a diheteroglycan, one containing three different monosaccharide units is a triheteroglycan, and so on. Generally diheteroglycans are either a linear chain of one single-unit branched with a second single-unit chain or a chain of polymers of alternating units.

3.3.3.1 Polysaccharide Solubility

Glycosyl unit of polysaccharides contains three hydroxyl groups on average. Thus each hydroxyl group is possible of hydrogen bonding to one or more water molecules. The ring oxygen atom and the glycosidic oxygen atom can also form hydrogen bonds with water. With every sugar unit in the chain having strong affinity for water, glycans possess great capacity to hold water molecules and readily hydrate, swell, and usually undergo partial or complete dissolution in aqueous systems. The mobility of water in food systems are modified by interaction with polysaccharides which in turn influences the physical and functional properties of polysaccharides. Many functional properties of foods, including texture, are controlled by joint action of polysaccharides and water.

The presence of polysaccharide molecules has sufficiently changed the structure of the water of hydration that is hydrogen-bonded to the polymer molecules so that it will not freeze. In some cases this water has also been referred to as plasticizing water. While motions of the hydration water molecules are retarded, they still can exchange freely and rapidly with other water molecules. In gels and fresh tissue the water of hydration accounts for only a fraction of the total water. Most other part of water is held in capillaries and cavities of variable sizes in the gel or tissue.

Rather than cryoprotectants polysaccharides function as cryostabilizers since their high-molecular-weight prohibits significant contribution to colligative properties such as increasing osmolality and depressing the freezing point of water. For example, when a starch solution is frozen, a two-phase system is formed, one of which is crystalline water (ice) and the other of which is a glass comprising about 70% starch molecules and 30% nonfreezable water. Occasionally polysaccharides provide cryostabilization by limiting ice crystal growth via adsorption to nuclei or active crystal growth sites.

Excluding very branch-on-branch structured molecules, most polysaccharides exist in some sort of helical shape. Certain highly ordered linear homoglycans, like cellulose, have flat, ribbon-like architectures. Such uniform linear chains assure strong intermolecular hydrogen bonding so as to produce crystallites divided by amorphous regions. Crystalline arrangements of this type are termed fringed micelles (Fig. 3.11). Such crystallites grant great strength, insolubility, and resistance to break down to cellulose fibers such as wood and cotton fibers. Owing to the crystalline regions cotton fibers are nearly inaccessible to enzyme penetration. However, crystallinity of most polysaccharides is not good enough to impart water insolubility, but are readily hydrated and dissolved.

Fig. 3.11 Fringed micelles. The crystalline regions are those in which the chains are parallel and ordered

Chain segments of most unbranched heteroglycans having nonuniform blocks of glycosyl units and most branched glycans cannot be closely packed over lengths necessary to form strong intermolecular bonding. Therefore they cannot form crystalline micelles and these molecules have a degree of solubility that increases as molecules are less able to fit together. In general, the solubility of polysaccharides is proportional to the degree of irregularity of the molecular chains.

3.3.3.2 Polysaccharide Solution Viscosity and Stability

Polysaccharides have remarkable ability to produce viscosity. They effectively adjust the flow properties and textures of beverage products and liquid food as well as the deformation properties of semisolid foods. They are generally used at concentrations of 0.25%-50%. A variety of factors including the size and shape of molecules and their conformations in the solvent determine the viscosity of a polysaccharide solution. The shapes of polysaccharide molecules in solution are regulated by the internal freedom around the glycosidic linkage bonds. The greater internal freedom, the more conformations are available. Thus such strong entropic drive provided by chain flexibility generally overcomes energy barrier leading to disordered or random coil states in solutions (Fig. 3.12). The specific nature (either compact or expanded) of the coils depends on the monosaccharide composition and linkages forming stiff coils not perfect random coils.

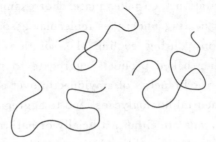

Fig. 3.12 Randomly coiled polysaccharide molecules

Viscosity depends on the shape (DP, molecular weight and the extension) and flexibility (rigidity) of the solvated polymer chain. Linear polymer molecules in solution sweep out a larger space than the branched ones of the same molecular weight (Fig. 3.13). They frequently collide with each other, generate friction and thereby produce viscosity. Even at low concentrations linear polysaccharides yield highly viscous solutions.

For highly branched molecules much less collision occurs and produces a much lower viscosity accordingly. This also implies that to produce the same viscosity at the same concentration significantly, larger molecules have to be applied in the case of highly branched polysaccharide. Likewise, linear polysaccharide chains carrying only uniform ionic charge increasing the end-to-end chain length because of repulsion of the like charges. Thus the polymer assuming an extended conformation sweeps out greater volume and produces solutions of high viscosity.

Molecular dispersions formed by dissolving unbranched, regular glycans in water under heating are unstable that will precipitate rapidly because segments of the long molecules collide and form intermolecular hydrogen bonds over the distance of a few units. Intermolecular associations then extend readily in a zipper-like fashion after initial short alignments. This organized nucleus will grow into the ordered, crystalline phase (fringed micelle) as other segments of other chains continue binding to it. Once a fringed micelle

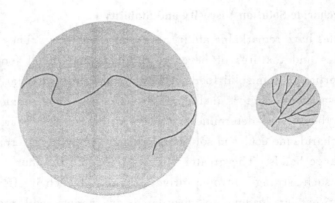

Fig. 3.13 Relative volumes occupied by a linear polysaccharide and a highly branched polysaccharide of the same molecular weight

reaches a certain size precipitation then takes place. For example, an aqueous solution of starch amylose prepared by heating undergoes molecular association and precipitation, a process called retrogradation, during cooling the solution. During cooling, amylose molecules in bread and other baked products aggregate to produce a firming. Staling occurs due to association of the branches of amylopectin over a longer storage time.

In general, all linear, neutral homoglycans tend to aggregate and partially crystallize. However, if unbranched glycans are either artificially or naturally derivatized, association of their segments was precluded and stable solutions result.

3.3.3.3 Polysaccharide Hydrolysis

Compared with proteins polysaccharides are relatively less stable to hydrolytic rupture and often experience depolymerization during food processing and/or storage. To eradicate undesirable viscosity at a relatively high concentration, food gums are deliberately depolymerized.

Hydrolysis of glycosidic bonds can be catalyzed by either acids or enzymes. A number of variables such as the acid strength, time, temperature, and structure of the polysaccharide determine the extent of depolymerization. Generally, many foods readily undergo hydrolysis during thermal processing because of their acidic nature. To compensate for defects associated with depolymerization during processing more of the polysaccharide or a more acid stable one is used in the formulation.

The rate and end products of enzymatic hydrolysis are governed by the specificity of the enzyme, pH, time, and temperature. It is due to the endogenous hydrolytic enzymes that polysaccharides are susceptible to microbial attack. This effect must be considered when gum products are employed as ingredients since they are very seldom delivered sterile.

3.3.4 Starch

As the predominant food reserve substance in plains, starch and modified starch products that are responsible for 70%-80% of human energy cost constitute most of the digestible carbohydrate in the human diet. Starches are provided by a vast variety of cereal grain seeds, tubers and roots. They together with their modifications have enormous food uses including adhesive, film forming, gelling, moisture retaining, stabilizing and texturizing applications.

Starch occurs naturally as discrete particles (granules) which are relatively dense and hydrate only slightly in cold water. Its thickening power is released only by heating a slurry of starch granules with stirring. Most starch granules are a mixture of two polymers: an essentially linear polysaccharide (amylose) and a highly branched polysaccharide (amylopectin).

3.3.4.1 Amylose

Usually amylose molecules bear molecular weights of about 10^6. While with virtually a linear chain of (1→4)-linked α-D-glucopyrarlosyl units, many amylose molecules have 0.3%-0.5% α-D-(1→6) branches of the linear linkages. The physical properties of amylose molecules are predominated by the linear backbones because the branch points are isolated by large distances.

The (1→4)-linkages of α-D-glucopyrarlosyl units assume a axial → equatorial configuration that delivers a helix conformation for amylase molecules (Fig. 3.14). The aliphatic groups located in the interior of the helix produce a lipophilic cavity, while the hydroxyl humps positioned on the exterior result in a hydrophilic shell.

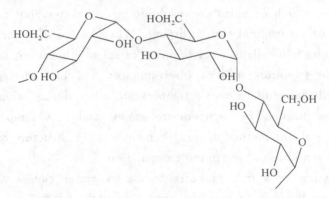

Fig. 3.14 A trisaccharide segment of an unbranched portion of an amylose or amylopectin molecule

3.3.4.2 Amylopectin

Amylopectin molecules typically having molecular weights of from 10^7 to 5×10^8 are among the largest molecules in nature. Highly branched amylopectin molecules are constituted with 4%-5% branch-point linkages. Amylopectin contains a chain with the only reducing end-group (C-chain), second-layer branches (B-chain), and third-layer branches (A-chain). These branches are clustered (Fig. 3.15) as double helices.

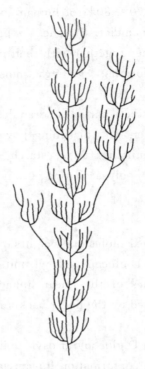

Fig. 3.15　A diagrammatic representation of a portion of an amylopectin molecule

3.3.4.3　Granule Gelatinization

Although intact starch granules are insoluble in cold water, but can swell slightly, and then resume to their original size on drying. When heated in water, molecular order within starch granules will collapse, such process called gelatinization, accompanying with irreversible granule swelling, loss of birefringence, and loss of crystallinity. Total gelatinization usually takes place over a temperature range during which larger granules generally gelatinize first. The temperature range and the temperature of initial gelatinization vary with the method of measurement and the structure of starch including granule type and distribution of the granule population.

Continued heating of starch granules in excess water causes additional granule swelling, continuous leaching of amylose, and eventually total disruption of granules, leading to a starch paste that is composed of a continuous phase (solubilized amylose and amylopectin) and a discontinuous phase (granule remnants). Usually complete molecular dispersion cannot be achieved unless high temperature, high shear and excess water are applied. It is highly unusual to use such conditions in food processing. When a hot cornstarch paste is cooled down, a viscoelastic and firm gel results.

3.3.4.4　Retrogradation and Staling

As mentioned before, a viscoelastic gel is produced once cooling a hot starch paste. In this process starch molecules form junction zones in the gel. Over the period of storage, starch molecules become progressively less soluble and will produce insoluble material which resist redissolving by heating. The whole process of dissolved starch becoming less soluble is termed as "retrogradation" which involving both amylose and amylopoctin. Amylose undergoes retrogradation much faster than amylopectin does. Several variables influence the rate of retrogradation including the molecular ratio of amylose to amylopoctin, structures of the amylose and amylopectin molecules, temperature, starch concentration, and other ingredients. Many food quality defects arise from starch retrogradation, such as bread staling and loss of viscosity in soups.

Staling starts as soon as the baked product begins to cool. In this process crumb firmness progressively increases along with a loss in product freshness. A number of factors including product formulation, the baking process, and storage conditions determine the rate of staling. The gradual transition of amorphous starch to a partially crystalline is the driving force for retrogradation. Association of outer branches is involved in retrogradation of amylopectin which requires a much longer time than amylose retrogradation.

3.3.5 Cellulose

As the most abundant carbohydrate on the earth cellulose, a high molecular-weight, linear, insoluble homopolymer, is composed of uniform β-D-glucopyranosyl units joined by (1→4) glycosidic linkages (Fig. 3.16). The linearity and stereoregularity of cellulose molecules warrant tight alignment over extended zones through hydrogen bonding, forming polycrystalline, and fibrous bundles. Amorphous regions separate and connect these crystalline regions. The strong hydrogen-bonding association render cellulose is insoluble since release of most of these hydrogen bonds at once is required for dissolution, but this can be achieved by substitution of cellulose.

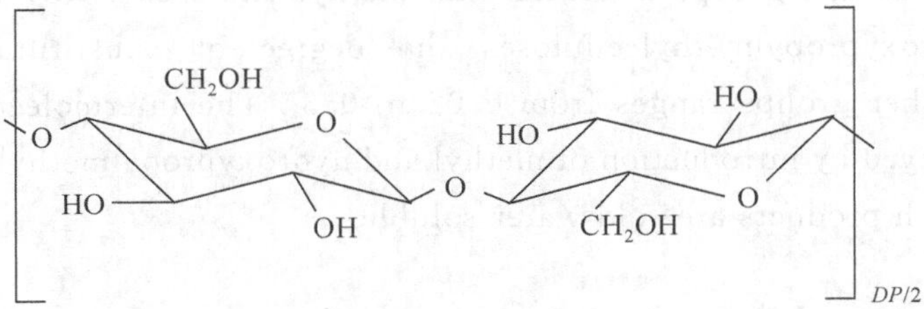

Fig. 3.16 Cellulose

3.3.5.1 Microcrystalline Cellulose

Microcrystalline cellulose (MCC) is widely used in the food industry which is prepared by hydrolysis of purified wood pulp, followed by separation of microcrystals of cellulose. Cellulose molecules typically of a chain of about 3,000, β-D-glucopyranosyl units align readily in long junction regions. However, the end of the crystalline region of cellulose chains diverges away from order into a more unorganized arrangement. During the process of hydrolysis of purified wood pulp with acid, the acid penetrates the lower density, amorphous zones, results in hydrolytic cleavage of chains in these regions, and consequentially releases individual crystallites. The chains of the released crystallites have greater freedom of motion, and can order themselves and grow larger.

3.3.5.2 Carboxymethylcellulose

Carboxymethylcellulose (CMC) is extensively employed as a food gum. To prepare CMC purified wood pulp is first treated with 18% sodium hydroxide solution and produces alkali cellulose. Then subjection of alkali cellulose to the sodium salt of chloroacetic acid delivers the sodium salt of the carboxymethyl ether (CMC). The degree of substitution (DS) for most commercial CMC products is in the range of 0.4-3.8. The most widely used type for a food ingredient bears a DS of 0.7.

Electrostatic repulsion exists in these fairly rigid molecules because they bear a negative charge that comes from numerous ionized carboxyl groups. This electrostatic repulsion forces these molecules assuming an extended conformation. Consequently, CMC solutions are both highly viscous and stable and the range of viscosity of CMC is broad and adjustable. Protein dispersions can be stabilized by CMC, and especially when it close to their isoelectric pH value. For example, CMC is applied to stabilize egg during co-drying or freezing, and casein precipitation in milk products is precluded through addition of CMC.

3.3.5.3 Methylcelluloses and Hydroxypropylmethylcelluloses

When alkali cellulose is treated with methyl chloride methyl groups (cellulose-O-CH_3) will be introduced. Hydroxypropylmethyl is another popular group for cellulose derivatization. Hydroxypropylmethylcelluloses (HPMC) are prepared by treatment of alkali cellulose with both propylene oxide and methyl chloride. The range of the degree of substitution with methyl groups in commercial methylcelluloses (MC) is 1.1 to 2.2. In commercial hydroxypropylmethylcelluloses the degree of substitution (MS) with hydroxypropyl ether groups ranges from 0.02 to 0.3. The intermolecuar association in cellulose is destroyed by introduction of methyl and hydroxypropylmethyl groups along the chains so that both products are cold-water soluble.

3.3.6 Pectin

Pectins are naturally present in the cell walls and intercellular layers of all land plants. These complex molecules are converted into commercial pectin products, galacturono-glycans with varying contents of methyl ester groups, by extraction with acid. Citrus peel and apple pomace are principal sources for commercial pectins. However, pectins from lemon peel are of the highest quality with advantage of the easiest isolation. Pectins are applied almost exclusively to form spreadable gels with the aid of sugar and acid or calcium ions.

The constitutions and properties of pectins depend on the source and preparation methods. In nature, around 80% of carboxyl groups in pectins are esterified with methyl groups. Partial hydrolysis of methyl ester groups as well as hydrolytic depolymerization occurs over extraction with acid. Therefore, the term "pectin" usually designates a family of water-soluble galacturonoglycans of variable contents of methyl ester groups and degrees of neutralization required for formation of gels. The free carboxylic acid groups may be partly or fully neutralized, typically in the sodium salt form.

Degree of esterification (DE) is defined as the ratio of esterified galacturonic acid groups to total galacturonic acid groups in pectins. Based on DE values pectins are classified in two classes: high methoxyl (HM) and low methoxyl (LM) pectins. Typical DE values for commercial HM-pectins range from 60% to 75% and those for LM-pectins range from 20% to 40%. HM-and LM-pectins gelatinize through different mechanisms. To achieve effective gelatinization, a minimum amount of soluble solids (i.e., 65% sugar content) and a pH within a narrow range (2.0-3.5) are required for HM-pectins. HM-pectin gels are thermally reversible. However, preparation of gels using LM-pectins requires a controlled amount of calcium or other divalent cations, but is independent of sugar content and not very sensitive to pH. Gelatinization of LM-pectins is usually faster than that of HM-pectins.

Glossary

acetal	乙缩醛
aldehyde	乙醛
aldose	醛糖
alignment	排成直线
aliphatic	脂肪族的
alkali	碱性的
ammonium ion	铵离子
amorphous	非结晶的
amylase	直链淀粉,多糖
amylopectin	支链淀粉
asymmetric	不对称的
beverage	饮料
biomolecule	生物分子
bisulfate	硫酸氢盐
calcium	钙

cane	甘蔗
caramel	焦糖
carbohydrates	碳水化合物
carboxymethylcellulose	羧甲基纤维素
catalyst	催化剂
cation	阳离子
cellulose	纤维素
chiral	手性的
chloroacetic	氯乙酸的
colloidal	胶状的
cryoprotectant	低温防护剂
dehydration	脱水
depolymerization	解聚合作用
dextrose	右旋糖
divalent	二价的
empirical formula	经验公式
epithelial	上皮的，皮膜的
equilibrium	平衡
eradicate	摧毁，根除
esterification	酯化作用
fermentation	发酵
flatulence	肠胃气胀
fragrance	芳香
furan	呋喃
furanose	呋喃糖
gelatinization	凝胶作用
gelatinize	成胶状
glucopyranose	吡喃型葡萄糖
glucose	葡萄糖
glycan	葡聚糖
glyceraldehyde	甘油醛
glycosidic bond	糖苷键
glycosylamine	葡基胺
granule	小颗粒
gum	树胶
hemiacetal	半缩醛
heterocyclic	杂环的
heteroglycan	杂多糖

hexose	己醛
homoglycan	同多糖
homopolymer	均聚物
humectants	湿润剂
hydrolyze	水解
hydroxylamine	羟胺
hydroxypropylmethylcellulose	羟丙基甲基纤维素
intestine	肠
isomer	同分异构体
juiciness	多汁性
ketohexose	己酮糖
ketone	酮
lactic acid	乳酸
lactose	乳糖
maltodextrin	麦芽糖糊精
maltol	麦芽糖醇
maltose	麦芽糖
metabolism	新陈代谢
methyl	甲基
methylcellulose	甲基纤维素
metric	公制的
molasse	糖浆
monosaccharide	单糖
mutagenic	诱导有机体突变的物质
nitrogen	氮
oligosaccharide	寡糖
organism	生物,有机体
osmolality	渗透度
pectin	胶质
phosphorus	磷
polarization	极化,产生极性
polymeric	聚合体的
polysaccharide	多糖
preservative	防腐剂
pyranose	吡喃糖
pyrolytic	热解的
raffinose	棉籽糖
reddish	微红的,淡红色的

FOOD CHEMISTRY

retrogradation	退化
rhamnose	鼠李糖
ribodesose	脱氧核糖
rupture	破裂,裂开
saccharide	糖
skeleton	骨架,框架
slurry	泥浆,浆
solubility	溶解性
stachyose	水苏糖
starch	淀粉
sucrose	蔗糖
sulfur	硫磺
syrup	糖浆
thermolysis	热分解作用
ubiquitous	普遍存在的
uronic acid	糖醛酸
viscosity	黏稠,黏性

Chapter 4 Protein

4.1 Introduction

4.1.1 Definition

Proteins are linear condensed products of various amino acids, which differ in molecular weight, electrical charge, and polar character. They are the source of essential and nonessential, dietary amino acids. At the elemental level, proteins contain 50%-55% carbon, 6%-7% hydrogen, 20%-23% oxygen, 12%-19% nitrogen, and 0.2%-3.0% sulfur.

All proteins are essentially consist of the same primary 20 amino acids; however, some proteins may not contain one or a few of these 20 amino acids. Because of the variety of side chains that occur when the amino acids are linked, different proteins may have widely different secondary and tertiary structures, in corresponding to the different chemical properties.

Proteins occur in animals as well as vegetable products in important quantities. In the developed countries, people intake more protein from animal products than plant ones. In other countries, the major portion of dietary protein is derived from plant products. Many plant proteins are lack of one or more of essential amino acids, and the protein content of some selected foods is listed in Table 4.1.

Table 4.1 Protein content of some selected foods

Product	Protein (g/100 g)
Meat: beef	16.5
pork	10.2
Chicken (light meat)	23.4
Fish: haddock	18.3
cod	17.6
Milk	3.6

(To be continued)

FOOD CHEMISTRY

(Table 4.1)

Product	Protein (g/100 g)
Egg	12.9
Wheat	13.3
Bread	8.7
Soybeans: dry, raw	34.1
cooked	11.0
Peas	6.3
Beans: dry, raw	22.3
cooked	7.8
Rice: white, raw	6.7
cooked	2.0
Cassava	1.6
Potato	2.0
Corn	10.0

4.1.2 Classification

Proteins are complex molecules, and the classification has been based mainly on solubility in different solvents. However, as more knowledge about molecular composition and structure is obtained, other criteria are being used for classification. These include behaviors in the ultracentrifuge and electrophoretic properties. Proteins are divided into the following main groups: simple, conjugated, and derived proteins.

4.1.2.1 Simple Proteins

Simple proteins yield only amino acids upon hydrolysis and include the following classes:

ⅰ) Albumins. Soluble in neutral, salt-free water. Usually these are proteins of relatively low molecular weight. Examples are egg albumin, lactalbumin, and serum albumin in the whey proteins of milk, leucosin of cereals, and legumelin in legume seeds.

ⅱ) Globulins. Soluble in neutral salt solutions and almost insoluble in water. Examples are serum globulins and lactoglobulin in milk, myosin and actin in meat, and glycinin in soybeans.

ⅲ) Glutelins. Soluble in very dilute acid or base and insoluble in neutral solvents. These proteins are common in cereals, such as glutenin in wheat and oryzenin in rice.

ⅳ) Prolamins. Soluble ethanol and insoluble in water. These proteins have large amounts of proline and glutamic acid and occur in cereals. Examples are zein in corn, gliadin in wheat, and hordein in barley.

ⅴ) Scleroproteins. Insoluble in water and neutral solvents and resistant to enzymic hydrolysis. These are fibrous proteins serving structural and binding purposes. Examples include elastin gelatin and keratin.

ⅵ) Histones. Basic proteins, as defined by their high content of lysine and arginine. Soluble in water and precipitated by ammonia.

ⅶ) Protamines. Strongly basic proteins of low molecular weight (4,000 to 8,000), rich in arginine. Examples are clupein and scombrin.

4.1.2.2 Conjugated Proteins

Conjugated proteins contain an amino acid part combined with a nonprotein material such as a lipid, nucleic acid, or carbohydrate. Some of the major conjugated proteins are as follows.

ⅰ) Phosphoproteins. An important group that includes many major food proteins. Phosphate groups are linked to the hydroxyl groups of serine and threonine. This group includes casein of milk and phosphoproteins of egg yolk.

ⅱ) Lipoproteins. They are combinations of lipids with protein and have excellent emulsifying capacity. Lipoproteins occur in milk and egg yolk.

ⅲ) Nucleoproteins. These are combinations of nucleic acids with protein. These compounds are found in cell nuclei.

ⅳ) Glycoproteins. These are combinations of carbohydrates with protein. Usually the amount of carbohydrate is small, but some glycoproteins have carbohydrate contents of 8 to 20 percent. An example of such a mucoprotein is ovomucin of egg white.

ⅴ) Chromoproteins. These are proteins with a colored prosthetic group. There are many compounds of this type, including hemoglobin and myoglobin, chlorophyll, and flavoproteins.

4.1.2.3 Derived Proteins

These are proteins obtained by chemical or enzymatic methods and divided into primary and secondary derivatives, depending on the extent of change that has taken place. Primary derivatives are slightly modified and are insoluble in water; an example is rennet coagulated casein. Secondary derivatives are more extensively changed and include proteoses, peptones, and peptides. The differences between these breakdown products lie in size and solubilities. All are soluble in water and not coagulated by heat.

4.2 Composition

4.2.1 Structure

Proteins are macromolecules with different levels of structural organization. The primary structure of proteins relates to peptide bonds between amino acids and also to the amino acid sequence in molecules.

The secondary structure of proteins involves folding the primary structure. Hydrogen bonds between amide nitrogen and carbonyl oxygen are the major stabilizing force. These bonds may be formed between different parts of the same polypeptide chain or between adjacent chains.

The tertiary structure of proteins involves a pattern of folding of the chains into a compact unit that is stabilized by hydrogen bonds, van der Waals forces, disulfide bridges, and hydrophobic interactions. The tertiary structure leads to the formation of a tightly packed unit with most of the polar amino acid residues located outside and hydrated.

Large molecules of molecular weights $> 50,000$ may form quaternary structures by association of subunits, stabilized by hydrogen bonds, disulfide bridges, and hydrophobic interactions.

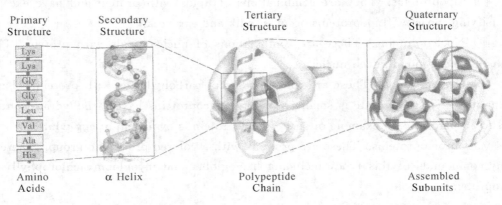

Fig. 4.1 Primary, secondary, tertiary and quaternary structures of proteins

4.2.2 Amino Acids

Amino acids, regarded as "building blocks" of protein, contain both basic amino and acidic carboxyl. α-amino acids are the basic structural units of proteins. These amino acids are composed of an α-carbon atom covalently attached to a hydrogen atom, an amino group, a carboxyl group, and a side chain R group.

Natural proteins contain up to 20 different primary amino acids, which are linked together via amide bonds (Table 4.2). These amino acids are different only because of the difference between the side chains. The physicochemical properties, such as net charge, solubility, chemical reactivity, and hydrogen bonding potential, of the amino acids are dependent on the chemical nature of the R group.

At neutral pH in aqueous solutions, both the amino and the carboxyl groups are ionized. The carboxyl group loses a proton and obtains a negative charge, while the amino group gains a proton and hence acquires a positive charge. Consequandy, amino acids possess dipolar characteristics.

Table 4.2 Amino acids (protein building blocks) with their corresponding three and one letter symbols

Amino Acid	Symbols	Amino Acid	Symbols	Amino Acid	Symbols
Glycine	(Gly, G)	L-Methionine	(Met, M)	L-Aspartic Acid	(Asp, D)
L-Alanine	(Ala, A)	L-Serine	(Ser, S)	L-Glutamic Acid	(Glu, E)
L-Valine	(Val, V)	L-Threonine	(Cys, C)	L-Lysine	(Lys, K)
L-Leucine	(Leu, L)	L-4-Hydroxyproline		L-5-Hydroxylysine	
L-Isoleucine	(Ite, I)	L-Tyrosine	(Tyr, Y)	L-Histidine	(His, H)
L-Proline	(Pro, P)	L-Asparagine[a]	(Asn, N)	L-Arginine	(Arg, R)
L-Phenylalanine	(Phe, F)	L-Glutamine[a]	(Gin, G)		
L-Tryptophan	(Trp, W)				

[a] When no distinction exists between the acid and its amide, the symbols (Asx, B) and (Glx, Z) are valid.

When the amino group of one amino acid reacts with the carboxyl group of another amino acid, a peptide bond is formed and a molecule of water is released. This C-N bond joins amino acids together to form proteins.

4.3 Properties

4.3.1 Denaturation

Denaturation is a process that changes the molecular structure without breaking any peptide bonds of a protein. The process is peculiar to proteins and affects different proteins to different degrees, depending on the structure of a protein. Therefore, ionizing radiation, shift in pH, change in temperature, or concentration of various ions, or addition of detergents or solvents, may cause dissociation of the oligomers into subunits, unfolding of the tertiary structure, and uncoiling of the secondary structure. Considering the complexity of many food systems, it is not surprising that denaturation is a complex process that cannot easily be explained in simple terms. Denaturation usually involves loss of biological activity and significant changes in some physical or functional properties such as solubility. The destruction of enzyme activity by heat is an important operation in food processing. In most cases denaturation is nonreversible; however, there are some exceptions, such as the recovery of some types of enzyme activity after heating. Heat denaturation is sometimes desirable, for example, the denaturation of whey proteins for the production of milk powder used in baking.

Denaturation may sometimes result in flocculation of globular proteins but may also lead to the formation of gels. Foods may be denatured, and their proteins destabilized, during freezing and frozen storage.

4.3.1.1 Physical Agents

Temperature and Denaturation

Heat is commonly used in food processing and preservation. Proteins undergo various degrees of denaturation during processing. This can affect their functional properties in foods, and thus it is important to understand the factors affecting protein denaturation.

Table 4.3 Thermal denaturation temperature (T_d) and mean hydrophobicities of proteins

Protein	T_d	Mean hydrophobi (kJ mol^{-1} residue^{-1})
Trypsinogen	55	3.68
Chymotrysinogen	57	3.78
Elastase	57	
Pepsinogen	60	4.02
Ribonuclease	62	3.24
Carboxyeptidase	63	
Alcoholdehydrogenase	64	
Bovine serum albumin	65	4.22
Hemoglobin	67	3.98
Lysozyme	72	3.72
Insulin	76	4.16
Eggalbumin	76	4.01
Trypsininhibitor	77	
Myoglobin	79	4.33
α-Lactalbumin	83	4.26
Cytochromec	83	4.26
β-Lactoglobulin	83	4.50
Avidin	85	3.81
Soyglycinin	92	
Oatglobulin	108	

When a protein solution is gradually heated above a critical temperature, it undergoes a sharp transition from the native to the denatured state. The temperature at the transition midpoint, where the concentration ratio of native and denatured states is 1, known as the denaturation temperature T_d. The mechanism of temperature-induced denaturation is highly complicated and involves primarily destabilization of the major noncovalent interactions. Hydrogen bonding, electrostatic, and van der Waals interactions are exothermic in nature. Therefore, they are destabilized at high temperatures and stabilized at low temperatures. However, as peptide hydrogen bonds in proteins are mostly buried in the interior, they remain stable over a wide range of temperature. On the other hand, hydrophobic interactions are endothermic, they are stabilized at high temperatures and destabilized at low temperatures. Therefore, as the temperature is increased, changes in the stabilities of these two groups of noncovalent interactions oppose each other.

However, the stability of hydrophobic interactions cannot increase infinitely with increasing temperature, because above a certain temperature, gradual breakdown of water structure will eventually destabilize hydrophobic interactions as well. The strength of hydrophobic interactions reaches a maximum at around 60~70 °C.

Another major force that affects conformational stability of proteins is the conformational entropy of the polypeptide chain. As temperature is increased, the increase in thermal kinetic energy of the polypeptide chain greatly facilitates unfolding of the polypeptide chain as temperature is increased. The temperature at which the sum of the free energies is zero is the denaturation temperature of the protein.

Hydrostatic Pressure and Denaturation

One of the thermodynamic variables that affects conformation of proteins is hydrostatic pressure. Unlike temperature-induced denaturation, which usually occurs in the range of 40~80 °C at 1 atmospheric pressure (atm); pressure-induced denaturation can occur at 25 °C if the pressure is great enough. Most proteins undergo pressure-induced denaturation in the range of 1-12 kbar as evidenced from changes in their spectral properties. The midpoint of pressure-induced transition occurs at 4-8 kbar.

Pressure induced denaturation of proteins occurs mainly because proteins are flexible and compressible. Although amino acid residues are densely packed in the interior of globular proteins, some void spaces invariably exist and this tends to be compressible.

Pressure-induced denaturation of globular proteins is usually accompanied by a reduction in volume of about 30-100 mL/mol. This decrease in volume is caused by two factors: elimination of void spaces as the protein unfolds, and hydration of the nonpolar amino acid residues that become exposed during unfolding. The later event results in a decrease in volume.

Pressure-induced protein denaturation is highly reversible. Most enzymes, in dilute solutions, regain their activity once the pressure is decreased. However, regeneration of near complete activity usually takes hours.

Shear and Denaturation

Denaturation of protein happens when high mechanical shear generated by shaking, whipping, kneading, etc. Many proteins denature and precipitate when they are vigorously agitated. In this circumstance denaturation occurs because of incorporation of air bubbles and adsorption of protein molecules to the air-liquid interface. Since the energy of the air-liquid interface is greater than that of the bulk phase, proteins undergo conformation changes at the interface. The extent of conformational change depends on the flexibility of the protein. Highly flexible proteins denature more readily at an air-liquid interface than do rigid proteins. The nonpolar residues of denatured protein orient toward the gas phase and the polar residues orient toward the aqueous phase.

Several food processing operations include high pressure, shear, and high temperature, for example, extrusion, high-speed blending, and homogenization. When a high shear rate is produced by a rotating blade, subsonic pulses are created and cavitation also occurs at the trailing edges of the blade. Both of these events can cause protein denaturation. The greater the shear rate, the greater the degree of denaturation. The combination of high temperature and high shear force leads to irreversible denaturation of proteins.

4.3.1.2 Chemical Agents

pH and Denaturation

Proteins are more stable against denaturation at the isoelectric point than any other pH. At neutral pH, most proteins are negatively charged, and only a few are positively charged. Since the net electrostatic repulsive energy is small compared to other favorable interactions, most proteins are stable at around neutral pH. However, at extreme pH values, strong intramolecular electrostatic repulsion caused by high net charge results in swelling and unfolding of the protein molecules. The degree of unfolding is greater at extreme alkaline pH values than it is at extreme acid pH values. The former behavior is attributed to ionization of partially buried phenolic, carboxyl, and sulfhydryl groups, which cause unraveling of the polypeptide chain as they attempt to be exposed to the aqueous environment. pH-induced denaturation is mostly reversible. However, in some cases, partial hydrolysis of peptide bonds, deamidation of Asn and Gln, destruction of sulfhydryl groups at alkaline pH, or aggregation can cause in irreversible denaturation of proteins.

Organic Solvents and Denaturation

Organic solvent can affect the stability of protein in different ways such as hydrophobic interactions, hydrogen bonding, and electrostatic interactions. Since nonpolar side chains are more soluble in organic solvents than in water, hydrophobic interactions are weakened by organic solvents. On the other hand, since the stability and formation of peptide hydrogen bonds are enhanced in a low-permittivity environment, certain organic solvents may actually promote the formation of peptide hydrogen bonds. The action of organic solvents on electrostatic interactions is two-fold. By decreasing permittivity, they enhance electrostatic interactions between oppositely charged groups and repulsion between groups with like charge as well. The net effect of an organic solvent on protein structure, therefore, usually depends on the magnitude of its effect on various polar and nonpolar interactions. At low concentration, some organic solvents can stabilize several enzymes against denaturation. At high concentrations, however, all organic solvents cause denaturation of proteins because of their solubilizing effect on nonpolar side chains.

Organic Solutes and Denaturation

Organic solutes, notably urea and guanidine hydrochloride (GuHCl), induce denaturation of proteins. For many globular proteins, the midpoint of transition from the native to denatured state occurs at 4-6 M urea and 3-4 M GuHCl at room temperature.

Complete transition often occurs in 8 M urea and in about 6 M GuHCl. GuHCl is a more powerful denaturant than urea because of its ionic character. Many globular proteins do not undergo complete denaturation even in 8 M urea, while in 8 M GuHCl they usually exist in a random coil state (completely denatured).

Denaturation of proteins by urea and GuHCl is believed to involve two mechanisms. The first mechanism involves preferential binding of urea and GuHCl to be denatured protein. Removal of denatured protein as a protein denaturant complex shifts the N→D equilibrium to the right. As the denaturant concentration is increased, continuous conversion of the protein to protein-denaturant complex eventually results in complete denaturation of the protein. Since binding of denaturant to denatured protein is very weak, a high concentration of denaturant is needed to cause complete denaturation. The second mechanism involves solubilization of hydrophobic amino acid residues in urea and GuHCl solutions. Since urea and GuHCl have the potential to form hydrogen bonds, at high concentration these solutes break down the hydrogen-bonded structure of water. This destructuring of solvent water makes it a better solvent for nonpolar residues. This results in unfolding mid solubilization of apolar residues from the interior of the protein molecule.

Urea or GuHCl-induced denaturation can be reversed by removing the denaturant. However, complete reversibility of protein denaturation by urea is sometimes difficult. This is because some urea converts to cyanate and ammonia. Cyanate reacts with amino groups and alters the charge of the protein.

Detergents and Denaturation

Detergents, such as sodium dodecyl sulfate (SDS) are powerful protein denaturing agents. Most of globular proteins will denature when SDS is at 3-8 mM concentration. The mechanism involves preferential binding of detergent to the denatured protein molecule. This causes a shift in equilibrium between the native and denatured states. Unlike urea and GuHCl, detergents bind strongly to denatured proteins, which is the reason that complete denaturation occurs at a relatively low detergent concentration of 3-8 mM. Because of this strong binding, detergent-induced denaturation is irreversible. Globular proteins denatured by SDS do not exist in a random coil state; instead, they assume an α-helical rod shape in SDS solutions. This rod shape is properly regarded as denatured.

Chaotropic Salts and Denaturation

Salts influence protein stability in two different ways. At low concentrations, ions interact with proteins via nonspecific electrostatic interactions. This electrostatic neutralization of protein charges usually stabilizes protein structure. Complete charge neutralization by ions occurs at below 0.2 ionic strength, and it is independent of the nature of the salt. However, at higher concentrations (>1 M), salts have ion specific effects that influence the structural stability of proteins. Salts such as Na_2SO_4 and NaF enhance, whereas NaSCN and $NaClO_4$ weaken it. Protein structure is influenced more by anions than by cations. At equal ionic strength, Na_2SO_4 and NaCl increase T_d, whereas NaSCN and $NaClO_4$ decrease it. Regardless of their chemical makeup and conformational differences, the structural stability of macromolecules, is adversely affected by high concentrations of salt. NaSCN and $NaClO_4$ are strong denaturants. The relative ability of various anions at isoionic strength to influence the structural stability of protein in general follows the series $F^- < SO_4^{2-} < Cl^- < Br^- < ClO_4^- < SCN^- < Cl_3CCOO^-$. This ranking is known as the Hofmeister series or chaotropic series. Fluoride, chloride, and sulfate salts are structure stabilizers, whereas the salts of other anions are structure destabilizers.

It is not well known how the salts can affect the structural stability of protein; however, their relative ability to bind and alter hydration properties of proteins is probably involved. Salts that stabilize proteins enhance hydration of proteins and bind weakly, whereas salts that destabilize proteins decrease protein hydration and bind strongly. These effects are primarily the consequence of energy perturbations at the protein-water interface. On a more fundamental level, protein stabilization or denaturation by salts is related to their effect on bulk water structure. Salts that stabilize protein structure also enhance the hydrogen-bonded structure of water, and salts that denature proteins also break down bulk water structure and make it a better solvent for polar molecules. In other words, the denaturing effect of chaotropic salts might be related to destabilization of hydrophobic interactions in proteins.

4.3.2 Gelation

A gel consists of a three-dimensional lattice of large molecules or aggregates, capable of immobilizing solvent, solutes, and filling materials. Protein gelation refers to transformation of a protein from the "sol" state to a "gel like" state. This transformation is facilitated by heat, enzymes, or divalent cations under appropriate conditions. All these agents induce the formation of a network structure. However, considering the types of covalent and noncovalent interactions involved, the mechanism of network formation can differ remarkably.

Most food protein gels are prepared by heating a protein solution. In this mode of gelation, the protein in a "sol" state is first transformed into a "progel" state by denaturation. The "progel" state is hormally a viscous liquid state in which some degree of protein polymerization has already occurred. This step causes unfolding of protein and exposure of a critical number of functional groups, such as hydrogen bonding and hydrophobic groups, so that the second stage, the formation of a protein network, can occur. Creation of the progel is irreversible because many protein-protein interactions occur between the unfolded molecules. When the progel is cooled to ambient or refrigeration temperature, the decrease in the thermal kinetic energy facilitates formation of stable noncovalent bonds among exposed the functional groups of the various molecules and thus leads to gelation.

The interactions involved in network formation are primarily hydrogen bonds, hydrophobic and electrostatic interactions. The relative contributions of these forces vary with types of protein, heating conditions, the extent of denaturation, and environmental conditions. Hydrogen bonding and hydrophobic interactions contribute more than electrostatic interactions to network formation except when multivalent ions are involved in cross linking. In general, some proteins carry a net charge, electrostatic repulsion occurs among protein molecules and this is not usually conducive to network formation. However, charged groups are essential for maintaining protein-water interactions and water holding capacity of gels.

Gel networks sustained primarily by noncovalent interactions are thermally reversible; that is, upon reheating they will melt to a progel state, as is commonly observed with gelatin gels. This is especially true when hydrogen bonds are the major contributor to the network. Since hydrophobic interactions are strong at elevated temperatures, gel networks formed by hydrophobic interactions are irreversible, such as egg-white gels. Proteins that contain both cysteine and cystine groups can undergo polymerization via sulfhydryl-disulfide interchange reactions during heating, and form a continuous covalent network upon cooling. Such gels are usually thermally irreversible. Examples of this type are ovalbumin, β-lactoglobulin, and whey protein gels.

Protein gels can be divided into two types: aggregated gels and clear gels, or coagulum (opaque) gels and translucent gels. Aggregated gels are formed from casein and from egg white proteins and are opaque because of the relatively large size of the protein aggregates. Clear gels are formed from smaller particles, such as those formed from whey protein isolate, and have high water-holding capacity. The type of gel formed by a protein depends on its molecular properties and solution conditions. Proteins containing large amounts of nonpolar amino acid residues undergo hydrophobic aggregation upon denaturation. These insoluble aggregates then randomly associate and set into an irreversible coagulum-type gel. Since the rate of aggregation and network formation is faster than the rate of denaturation, proteins of this type readily form a gel network even while being heated. The opaqueness of these gels is due to light scattering caused by the unordered network of insoluble protein aggregates.

Proteins that contain small amounts of nonpolar amino acid residues form soluble complexes upon denaturation. Since the rate of association of these soluble complexes is slower than the rate of denaturation, and the gel network is predominantly formed by hydrogen bonding interactions, they often do not set into a gel until heating followed cooling has occurred (8%-12% protein concentration assumed). Upon cooling, the slow rate of association of the soluble complexes facilitates formation of an ordered translucent gel network.

The stability of a gel network against thermal and mechanical forces is dependent on the number of types of cross-links formed per monomer chain. Thermodynamically, a gel network would be stable only when the sum of the interaction energies of a monomer in the gel network is greater than its thermal kinetic energy. This is dependent on several intrinsic (such as the size, net charge) and extrinsic (such as pH, temperature, and ionic strength) factors. The square root of the hardness of protein gels exhibits a linear relationship with molecular weight. Globular protein with molecular weight $<23,000$ can form a heat-induced gel only when they contain at least one free sulfhydryl group or a disulfide bond. The sulfhydryl groups and disulfide bonds facilitate polymerization, and thus increasing the effective molecular weight of polypeptides to $> 23,000$. Gelatin preparations with effective molecular weights of less than 20,000 cannot form a gel.

Another critical factor is protein concentration. In order to form a self standing gel network, a minimum protein concentration, known as least concentration endpoint (LCE), is required. The LCE is 8% for soy proteins, 3% for egg albumin, and about 0.6% for gelatin.

Several environmental factors, such as pH, salts, and other additives, may also affect gelation of proteins. At or near isoelectric pH, proteins usually form coagulum-type gels. At extremes of pH, weak gels are formed because of strong electrostatic repulsion. The optimum pH for gel formation is about 7-8 for most of proteins.

Formation of protein gels can sometimes be facilitated by limited proteolysis. A well-known example is cheese. Addition of chymosin (rennin) to casein micelles in milk results in the formation of a coagulum type gel. This is achieved by cleavage of kapa-casein, a micelle component, causing release of a hydrophilic portion, known as the glycomacropeptide. The remaining so-called para-casein micelles possess a highly hydrophobic surface that facilitates formation of a weak gel network.

Divalent cations, such as Ca^{2+} and Mg^{2+}, can also be used to form protein gels. These ions form cross-links between negatively charged groups of protein molecules. A good example of this type of gel is tofu from soy proteins. Alginate gels also can be formed by this way.

4.3.3 Emulsifying Properties

Several natural and processed foods, such as milk, egg yolk, coconut milk, soy milk, butter, sausage, and cakes, are emulsion-type products where proteins play an important role as emulsifiers. In natural milk, fat globules are stabilized by a membrane composed of lipoproteins. When milk is homogenized, the lipoprotein membrane is replaced by a protein film comprised of casein micelles and whey proteins. Homogenized milk is more stable against creaming than natural milk because the casein micelle-whey protein film is stronger than the natural lipoprotein membrane.

Several procedures are used to determine the efficiency of proteins in emulsifying lipids and the stability, such as size distribution of oil droplets formed, emulsifying activity, emulsion capacity, and emulsion stability.

4.3.3.1 Emulsifying Activity Index

The physical and sensory properties of a protein-stabilized emulsion depend on the size of the droplets formed and the total interfacial area created.

The average droplet size of emulsions can be determined by several methods, such as light microscopy, electron microscopy, light scattering, or use of a Coulter counter.

4.3.3.2 Protein Load

The amount of protein adsorbed at the oil-water interface of an emulsion has a bearing on its stability. To determine the amount of protein adsorbed, the emulsion is centrifuged, the aqueous phase is separated, and the cream phase is repeatedly washed and centrifuged to remove any loosely adsorbed proteins. The amount of protein adsorbed to the emulsion particles is determined from the difference between the total protein initially present in the emulsion and the amount present in the wash fluid from the cream phase. Knowing the total interfacial area of the emulsion particles, the amount of protein adsorbed per square meter of the interfacial area can be calculated.

As the volume fraction of the oil phase increases the protein load decreases at constant protein content in the total emulsion. For high-fat emulsions and small sized droplets, more protein is obviously needed to adequately coat the interfacial area and stabilize the emulsion.

4.3.3.3 Emulsion Capacity

Emulsion capacity (EC) is represented by the volume of oil (cm^3) that is emulsified in a model system by 1 g of protein when oil is added continuously to a stirred aliquot of solution or dispersion. This method involves addition of oil or melted fat at a constant rate and temperature to an aqueous protein solution that is continuously agitated in a food blender. Phase inversion is detected by a sudden change in viscosity or color (usually a dye is added to the oil), or by an increase in electrical resistance. Inversion is not instantaneous, but is preceded by formation of a water-in-oil-in-water double emulsion. Since emulsion capacity is expressed as volume of oil emulsified per gram protein at phase inversion, it decreases with increasing protein concentration once a point is reached where unabsorbed protein accumulates in the aqueous phase. Therefore, to compare emulsion capacities of different proteins, EC versus protein concentration profiles should be used instead of EC at a specific protein concentration.

4.3.3.4 Emulsion Stability

Protein-stabilized emulsions are often stable for days. Thus, a detectable amount of creaming or phase separation is not often observed in a reasonable time period when samples are stored at atmospheric conditions. Therefore, drastic conditions, such as storage at elevated temperature or separation under centrifugal force, is often used to evaluate emulsion stability. If centrifugation is used, stability is then expressed as percent decrease in interfacial area (i.e., turbidity) of the emulsion, or percent volume of cream separated, or as the fat content of the cream layer. Normally, emulsion stability is expressed as

$$ES = \text{volume of cream layer} / \text{total volume of emulsion} \times 100$$

where the volume of the cream layer is measured after a standardized centrifugation treatment. A common centrifugation technique involves centrifugation of a known volume of emulsion in a graduated centrifuge tube at $1300 \times g$ for 5 min. The volume of the separated cream phase is then measured and expressed as a percentage of the total volume. Sometimes centrifugation at a relatively low gravitational force ($180 \times g$) for a longer time (15 min) is used to avoid coalescence of droplets.

4.3.3.5 Factors Influencing Emulsification

The properties of protein-stabilized emulsions are affected by some factors. These include intrinsic factors, such as pH, ionic strength, temperature, presence of low-molecular-weight surfactants, sugars, oil-phase volume, type of protein, and the melting point of the oil used; and extrinsic factors, such as type of equipment, rate of energy input, and rate of shear.

Solubility plays a role in emulsifying properties, but 100% solubility is not an absolute requirement. While highly insoluble proteins do not perform well as an emulsifiers, no reliable relationship exists between solubility and emulsifying properties in the 25%-80% solubility range. However, since the stability of a protein film at the oil-water interface is dependent on favorable interactions with both oil and aqueous phases, some degree of solubility is likely to be necessary. The minimum solubility requirement for good performance may vary among proteins.

The pH of the environment affects the emulsifying properties by changing the solubility and surface hydrophobicity of proteins, as well as the charge of the protective layer around the lipid globules.

The emulsifying properties of proteins show a weak positive correlation with surface hydrophobicity, but not with mean hydrophobicity. The ability of various proteins to decrease interfacial tension at the oil-water interface and to increase the emulsifying activity index is related to their surface hydrophobicity values. However, this relationship is by no means perfect. The emulsifying properties of several proteins, such as β-lactoglobulin, α-lactalbumin, and soy proteins, do not show a strong correlation with their surface hydrophobicity.

Partial denaturation of proteins prior to emulsification, which does not lead to insolublization, usually improves the emulsifying properties. This is due to increased molecular flexibility and surface hydrophobicity. The rate of unfolding at an interface depends on the flexibility of the original molecule. In the unfolded state, proteins containing free sulfhydryl groups and disulfide bonds undergo slow polymerization via disulfide-sulfhydryl interchange reaction. This causes to formation of a highly viscoelastic film at the oil-water interface. Heat denaturation that is sufficient to cause insolublization impairs emulsifying properties of proteins.

4.3.4 Foaming Properties

Food foams are dispersions of gas bubbles in a continuous liquid or semisolid phase. Many processed foods are foam-type products, including, ice cream, cakes, bread, souffles, mousses, and marshmallow. The unique textural properties and mouthfeel of these products come from the dispersed tiny air bubbles. In most of these products, proteins are the main surface-active agents that contribute to the formation and stabilization of the dispersed gas phase.

Generally, protein stabilized foams are formed by bubbling, whipping or shaking a protein solution. The foaming property of a protein refers to its ability to form a thin tenacious film at gas-liquid interfaces so that large quantities of gas bubbles can be incorporated and stabilized. Foaming properties are evaluated by several ways. The foamability or foaming capacity of a protein refers to the amount of interfacial area that can be created by the protein. It can be expressed in several ways, such as overrun (or steady-state foam value), or foaming power (or foam expansion).

The foaming power, FP, is expressed as

$$FP = \frac{\text{volume of gas incorporaed}}{\text{volume of liquild}} \times 100$$

$$\text{The steady-state foam value} = \frac{\text{volume of foam}}{\text{volume of initial liquid phase}} \times 100$$

$$\text{Overrun} = \frac{\text{volume of foam} - \text{volume of initial liquid}}{\text{initial liquid volume of initial liquid}} \times 100$$

Foaming power generally increases with protein concentration until a maximum value is reached. It is also affected by the method used for the foam formation. FP at a given protein concentration is often used as a basis for comparing the foaming properties of various proteins. The foaming powers of various proteins at pH 8.0 are given in Table 4.3.

Table 4.3 Comparative foaming power of protein solutions

Protein type	Foaming power at 0.5% protein concentration (w/v)
Bovine serum albumin	280%
Whey protein isolate	600%
Egg albumen	240%
Ovalbumin	40%
Bovine plasma	260%
β-lactoglobulin	480%
Fibrinogen	360%
Soy protein (enzyme hydrolyzed)	500%
Gelatin (acid processed pigskin)	760%

Foam stability refers to the ability of protein to stabilize foams against gravitational and mechanical stresses. Foam stability is often expressed as the time required for 50% of the liquid to drain from a foam or for a 50% reduction in foam volume. These are very empirical methods, and they do not provide fundamental information about factors concerning foam stability.

4.4 Food Proteins

4.4.1 Animal Proteins

4.4.1.1 Milk Proteins

Milk proteins have found various applications in formulated foods, such as meat extenders. The proteins of cow's milk can be divided into two groups: caseins, which are phosphoproteins and comprise 78 percent of the total weight, and milk serum proteins, which make up 17 percent of the total weight. The latter group includes β-lactoglobulin (8.5 percent), α-lactalbumin (5.1 percent), immune globulins (1.7 percent), and serum albumins. In addition, about 5 percent of milk's total weight is nonprotein nitrogen (NPN)-containing substances, which include peptides and amino acids. Milk also contains very small amounts of enzymes, including peroxidase, acid phosphatase, alkaline phosphatase, xanthine oxidase, and amylase. The protein composition of bovine milk is listed in Table 4.4.

Table 4.4 Protein composition of bovine milk

Protein	% of Total Protein
Caseins	80
α-caseins	42
β-caseins	25
κ-caseins	9
Whey proteins	20
α-lactalbumin	4
β-lactoglobulin	9

Casein is defined as the heterogeneous group of phosphoproteins precipitated from skim milk at pH 4.6 and 20 ℃. The proteins remaining in solution, the serum or whey proteins can be separated into the classic lactoglobulin and lactalbumin fractions by half saturation with ammonium sulfate or by full saturation with magnesium sulfate.

However, this separation is possible only with unheated milk. After heating by boiling, for instance, approximately 80 percent of whey proteins will precipitate with the casein at pH 4.6; this property has been used to develop a method for measuring the degree of heat exposure of milk and milk products.

Casein exists in milk as relatively large, nearly spherical particles of 30 to 300 nm in diameter. In addition to acid precipitation, casein can be separated from milk by rennet action or by saturation with sodium chloride. The composition of the casein depends on the method of isolation. In the native state, caseinate particles contain relatively large amounts of calcium and phosphorus, smaller quantities of magnesium, and citrate, so that they are usually referred to as calcium caseinatephosphate or calcium phosphocaseinate particles. When adding acid to milk, calcium and phosphorus are progressively removed until, at the isoelectric point of pH 4.6, then the casein is completely free of salts. Other methods of treating casein yield other products; for example: salt precipitation does not remove the calcium and phosphorus, and rennet action involves limited proteolysis. The rennet casein is named paracasein.

Casein is a nonhomogeneous protein that consists of three types, identified as α-($\alpha s1$- and $\alpha s2$-), β- and κ-casein. One part of the α-casein is precipitable by calcium ions and has been designated calcium-sensitive casein or αs. The non-calcium-sensitive fraction, κ-casein, is assumed to confer stability on the casein micelle; this has been found to be removed by the action of rennin, thereby leaving the remaining casein perceptible by calcium ions. κ-casein is the fraction with the lowest phosphate content. The two αs-caseins show the strong association. The association of β-casein is temperature dependent. At 4 °C only monomers exist; at temperatures greater than 8 °C association will occur. $\alpha s1$-casein has more acidic than basic amino acids and has a net negative charge of 22 at pH 6.5. The polypeptide chain contains 8.5 percent proline that is distributed uniformly, resulting in no apparent secondary structure. $\alpha s2$-casein has the highest amounts of phosphorylations and a low proline content. β-casein is a single polypeptide chain with a total of 209 amino acids. The distribution of amino acids in the polypeptide chain is quite specific. The N-terminal segment has a high negative charge, giving it hydrophilic properties; the C-terminal portion is highly hydrophobic. This arrangement leads surfactant properties to the protein.

It appears that micelles are formed by cross-linking of some ester phosphate groups by calcium. Chelation of calcium results in dissociation and solubilization of micelles, and the rate of chelation corresponds to the ester phosphate contents of the monomers.

The whey proteins of milk were originally thought to be composed of two main components, lactalbumin and lactoglobulin. Then these were found that the lactalbumin contains a protein with the characteristics of a globulin. This protein, known as β-lactoglobulin, is the most abundant among the whey proteins. It has a molecular weight of 36,000. In addition to β-lactoglobulin, the classic lactalbumin fraction contains α-lactalbumin, serum albumin, and at least two minor components.

β-lactoglobulin is rich in lysine, leucine, glutamic acid, and aspartic acid. It is a globular protein β-lactoglobulin has a tightly packed structure and consists of eight strands of antiparallel β sheets. The interior of the molecule is hydrophobic. The molecular structure also contains a certain amount of α helix, which plays a role in the formation of the usually occurring dimer. The association is pH dependent. β-lactoglobulin will form octamers at low temperature, and high concentration and at pH values between 3.5 and 5.2. Below pH 3.5, the protein dissociates into monomers. This protein is the only milk protein containing cysteine, therefore, contains free sulfhydryl groups, which play a role in the development of cooked flavor in heated milk. The cysteine group is also involved in the thermal denaturation. At pH 6.7 and above 67 ℃, β-lactoglohulin denatures, followed by the aggregation. The first step in the denaturation is a series of reversible conformational changes that cause exposure of cysteine. The next step involves association through the sulfhydryl-disulfide exchange.

The amino acid sequence of α-lactalbumin is very similar to that of hen egg-white lysozyme. α-lactalbumin has a high binding capacity for calcium and some other metals. It is insoluble at the isoelectric range from pH 4 to 5. The calcium in α-lactalbumin is bound very strongly and protects the stability of the molecule against the thermal denaturation.

Whey protein concentrate is made from whey, which is the by-product of the cheese making. Removal of lactose and minerals requires the reverse osmosis end ultrafiltration processing.

4.4.1.2 Meat Proteins

The proteins of muscle are composed of about 70 percent structural or fibrillar proteins and about 30 percent water-soluble proteins. The fibrillar proteins contain about 32 to 38 percent myosin, 13 to 17 percent actin, 7 percent tropomyosin, and 6 percent stroma proteins. Meat and fish proteins contribute to highly organized structures that lend particular properties to these products. Some of the other proteins discussed in this chapter are more or less globular and consist of particles that are not normally involved in an extensive structural array. Examples are milk proteins and proteins in cereals and oilseeds. Extensive structure formation involving these proteins may occur in various technological processes such as making cheese from milk or texturized vegetable protein products from oilseeds.

Meat contains three general types of proteins: soluble proteins, which can easily be removed by extraction with weak salt solutions (ionic strength $\leqslant 0.1$); contractile proteins; and stroma proteins of the connective tissue. The soluble proteins are classed as myogens and myoalbumins. The myogens are a heterogeneous group of metabolic enzymes. After extraction of the soluble proteins, the fibril and stroma proteins remain. They can be extracted with buffered 0.6 M potassium chloride to yield a viscous gel of actomyosin.

Myosin is the most abundant muscle proteins and accounts for about 38 percent of the total. Myosin is a highly asymmetric molecule with a molecular weight of about 500,000 that contains approximately 60 to 70 percent α-helix structures. The molecule has a relatively high charge and contains large amounts of glutamic and aspartic acids and dibasic amino acids. Myosin has enzyme activity and can split ATP into ADP and monophosphate, thereby liberating energy that is used in the muscle contraction. The myosin molecule is not a single entity. It can be separated into two subunits by means of enzymes or the ultracentrifuge. The subunits with the higher molecular weight about 220,000, are called heavy meromyosin. Those with low molecular weight, about 20,000, are called light meromyosin. Only the heavy meromyosin has ATP-ase activity.

Actin makes up about 13 percent of the muscle protein, so the actin-myosin ratio is about 1 : 3. Actin occurs in two forms: G-actin and F-actin (G and F denote globular and fibrous). G-actin is a monomer that has a molecular weight of about 47,000 and is a molecule of almost spherical shape. Because of its relatively high proline content, it has only about 30 percent of α-helix configuration. F-actin is a large polymer and is formed when ATP is split from G-actin. The traits of actin combine to form a double helix of indefinite length, and molecular weights of actin have been reported to be in the order of several millions.

Actomyosin is a complex of F-actin and myosin is responsible for muscle contraction and relaxation. Contraction occurs when myosin ATP-ase activity splits ATP to form phosphorylated actin and ADP. For this reaction, the presence of K^+ and Mg^{2+} is required. Relaxation of muscle depends on regeneration of ATP from ADP by phosphorylation from creatine phosphate.

Collagen

The contractile meat proteins are separated and surrounded by layers of connective tissues. The amount and nature of this connective tissue is an important factor for the tenderness or toughness meat. Collagens form the most widely occurring group of proteins in the animal body. They are part of the connective tissues in muscles and organs, skins, bones, tooth, and tendons. Collagens are a distinct class of proteins that can be demonstrated by the X-ray diffraction analysis. This technique shows that collagen fibrils have the regular periodicity of 64 nm, which can be increased under tension to 400 nm. Collagen exists as a triple helix. The triple helix is the tropocollagen molecule; these are lined up in a staggered array, overlapping by one-quarter of their length to form a fibril. The fibrils are stacked in layers to form connective tissues. Important in the formation of these structures is the high content of hydroxyproline and hydroxylysine. The content of dibasic and diacidic amino acids is also high, but tryptophan and cystine are absent. As a result of this particular amino acid composition, there are few interchain cross-bonds, and collagen swells readily in acid or alkali.

Heating of collagen fibers in water to 60-70°C shortens them by one-third or one-fourth of the original length. This temperature is characteristic of the type of collagen and is called the shrink temperature (Ts). The Ts of fish skin collagen is very low, 35 ℃. When the temperature is increased to about 80 ℃, the mammalian collagen transforms into gelatin. Certain amino acid sequences are common in collagen, such as Ghy-Pro-Hypro-Gly. In a triple helix, only certain sequences are permissible. The structural unit of the collagen fibrils is tropocollagen with a length of 280 nm, a diameter of 1.5 nm, and a molecular weight of 360,000. Gelatin is a soluble protein made from the insoluble collagen. The process of transforming collagen into gelatin involves the following three changes:

ⅰ. Rupture of a limited number of peptide bonds to reduce the length of chains;
ⅱ. Rupture or disorganization of a number of the lateral bonds between chains;
ⅲ. Changes in the chain configuration.

The last change is the only change that is essential for the conversion of collagen to gelatin. The conditions used during the production of gelatin determine its characteristics. If there are extensive breakdown of peptide bonds, many lateral bonds may remain intact and soluble fragments are produced. If many lateral bonds are destroyed, gelatin molecules may have relatively long chain lengths. Thus, there is a great variety of gelatins. In normal productions, hides or bones are extracted first, under relatively mild conditions, followed by successive extractions under more severe conditions. The first extraction yields the best-quality gelatin. The term gelatin is used for products derived from mammalian collagen that can be dispersed in water and show a reversible sol-gel change with temperature. The gels formed by gelatin can be considered as a partial return of molecules to an ordered state. However, the return to the highly ordered state of collagen is not possible. High-quality gelatin has an average chain length of 60,000 to 80,000, whereas the value for native collagen is infinite.

Fish Proteins

The proteins of fish flesh can be divided into three groups based on solubility. The skeletal muscle of fish consists of short fibers arranged between sheets of connective tissues, although the amount of connective tissues in fish muscle is less than that in mammalian tissues and the fibers are shorter. The myofibrils of fish muscle have a striated appearance similar to that of mammalian muscle and contain the same major proteins, myosin, actin, actomyosin, and tropomyosin. The soluble proteins include most of the muscle enzymes and account for about 22 percent of the total protein. The connective tissue of fish muscle is present in lower quantity than in mammalian muscle; the tissue has different physical properties, which result in a more tender texture of fish, compared with meat. The structural proteins consist mainly of actin and myosin, and actomyosin accounts about three-quarters of the total muscle protein. Fish actomyosin has been found to be quite labile and easily changed during the processing and the storage. During the frozen storage, the actomyosin becomes progressively less soluble, and the flesh becomes increasingly tough.

Egg Proteins

The proteins of eggs are characterized by their high biological value and can be divided into the egg white and egg yolk proteins. Egg albumen is used in various food formulations because of its foaming properties and heat-gelling ability, while egg yolk serves as an emulsifying agent. The egg white contains at least eight different proteins, which are listed in Table 4.5. Some of these proteins have unusual properties, as shown in Table 4.5; for example, lysozyme is an antibiotic, ovoimucoid is a trypsin inhibitor, ovomucin inhibits hemagglutination, avidin binds biotin, and conalbumin binds iron. The antimicrobial properties help to protect the egg from the bacterial invasion.

Table 4.5 Protein composition of egg white

Constituent	Approximate amount (%)	Approximate isoelectric Point (pH)	Unique properties
Ovalbumin	54	4.6	Denatures easily, has sulfhydryls
Conalbumin	13	6.0	Complexes iron, antimicrobial
Ovomucoid	11	4.3	Inhibits enzyme trypsin
Lysozyme	3.5	10.7	Enzyme for polysaccharides antimicrobial
Ovomucin	3.5	4.5-5.0	Viscous, high sialic acid, reacts with viruses
Flavoprotein-apoprotein	0.8	4.1	Binds riboflavin
Proteinase inhibitor	0.1	5.2	Inhibits enzyme (bacterial proteinase)
Avidin	0.05	9.5	Binds biotin, antimicrobial
Unidentified proteins	8	5.5, 7.5	Mainly globulins
Nonprotein	8	8.0, 9.0	Primarily half glucose and salts (poorly characterized)

Egg yolk proteins precipitate when the yolk is diluted with water. The protein components of egg yolk are listed in Table 4.6. The yolk contains a large amount of lipid, part of which occurs in bound form as lipoproteins. Lipoproteins are excellent emulsifiers, so egg yolk is widely used in foods. The two lipoproteins are lipovitellins, which has 17 to 18 percent lipid, and lipovitellenins, which has 36 to 41 percent lipid. The protein portions of these compounds after removal of the lipid are named vitellin and vitellenin. The former contains 1 percent phosphorus, while the latter 0.29 percent.

Table 4.6 Protein components of egg yolk

Constituent	Approximate amount(%)	Particular properties
Livetin	5	Contains enzymes—poorly characterized
Phosvitin	7	Contains 10% phosphorus
Lipoproteins	21	Emulsifiers
(Total protein)	(33)	

When fluid egg yolk is frozen, changes take place, causing the thawed yolk to form a gel. Gelation increases as the freezing temperature is lowered from $-6℃$ to $-14℃$. Gradual aggregation of lipoprotein is hypothesized as the cause of gelation.

4.4.2 Plant Proteins

4.4.2.1 Soybean Proteins

Soybean proteins are used in a variety of traditional products, such as soymilk, defatted flour, grits. The proteins in soybeans are contained in protein bodies, or aleurone grains, which measure from 2 to 20 gins in diameter. The protein bodies can be visualized by the electron microscopy. Soy protein is a good source of all the essential amino acids except methionine and tryptophan. The high lysine content makes it a good complement to cereal proteins, which are low in lysine. Soybean proteins have neither gliadin nor glutenin, the unique proteins of wheat gluten. As a result, soy flour cannot be incorporated into bread without the use of special additives that improve loaf volume. The soy proteins have a relatively high solubility in water or dilute salt solutions at pH values below or above the isoelectric point. This means they can be classified as globulins.

Application of heat to soybeans or defatted soy meal makes the protein progressively more insoluble. Hydrogen bonds and hydrophobic bonds appear to be responsible for the decrease in solubility of the proteins during heating.

4.4.2.2 Wheat Proteins

Wheat proteins are unique among plant proteins and are responsible for bread-making properties. The classic method of fractionation based on solubility characteristics indicates the presence of four main fractions: albumin, which is water-soluble and coagulated by heat; globulin, soluble in the neutral salt solution; gliadin, a prolamine soluble in the 70 percent ethanol; and glutenin, a glutelin insoluble in alcohol but soluble in dilute acid or alkali.

Gliadin and glutenin are the gluten-forming proteins of wheat. The formation of gluten occurs when flour is mixed with water. The gluten is an important agent which can provide the basis for the crumb structure of bread, because it is a coherent elastic mass, which holds together other bread components such as starch and gas bubbles. The hydration of the gluten proteins results in the formation of fibrils, with gliadins forming films and glutenins forming strands.

Gluten proteins have a high content of glutamine but low in the essential amino acids lysine, methionine, and tryptophan. The insolubility of gluten proteins can be directly related to their amino acid compositions. High levels of nonpolar side chains result from the presence of glutamic and aspartic acids as the amides. Because these are not ionized, there is a high level of hydrogen bonding, which contributes to aggregation of molecules and results in low solubility.

FOOD CHEMISTRY

Heat damage to gluten can result from excessive air temperatures used in the drying of wet grains. The gluten becomes tough and is more difficult to extract. Therefore, heat-denaturated wheat gives bread poor textures and loaf volumes.

4.5 Peptides

4.5.1 Properties

Peptides are formed by binding amino acids together through an amide linkage. On the other hand, peptide hydrolysis results in free amino acids.

Peptides are denoted by the number of amino acid residues as di-, tri-, tetrapeptides, etc., and the term "oligopeptides" is used for those with 10 or less amino acid residues. Higher molecular weight peptides are called polypeptides.

The pK values and isoelectric points for some peptides are listed in Table 4.7. The acidity of the free carboxyl groups and the basicity of the free amino groups are lower in peptides than in the corresponding free amino acids. The amino acid sequence also has an effect (e.g., Gly-Asp/Asp-Gly).

Table 4.7 Dissociation constants and isoelectric points of various peptides (25 ℃)

Peptide	pK_1	pK_2	pK_3	pK_4	pK_5	pI
Gly-Gly	3.12	8.17				5.65
Gly-Gly-Gly	3.26	7.91				5.59
Ala-Ala	3.30	8.14				5.72
Gly-Asp	2.81	4.45	8.60			3.63
Asp-Gly	2.10	4.53	9.07			3.31
Asp-Asp	2.70	3.40	4.70	8.26		3.04
Lys-Ala	3.22	7.62	10.70			9.16
Ala-Lys-Ala	3.15	7.65	10.30			8.98
Lys-Lys	3.01	7.53	10.05	11.01		10.53
Lys-Lys-Lys	3.08	7.34	9.80	10.54	11.32	10.93
Lys-Glu	2.93	4.47	7.75	10.50		6.10
His-His	2.25	5.60	6.80	7.80		7.30

4.5.2 Bioactive Peptides in Food

4.5.2.1 Casein Phosphopeptide

Since casein contains some clusters of phosphorylated serine residues several phosphopeptides were liberated by enzymatic *in-vitro*-proteolysis as well as during intestinal digestion of caseins. After ingestion of a casein-containing diet, it was possible to isolate the α-casein fragment from the soluble part of intestinal chime.

Casein phosphopeptides that are generated by *in-vitro* proteolysis of casein using dissolved and immobilized trypsin, have been characterized in terms of their potential nutritive and functional relevance by examining their abilities to bind calciums and irons.

4.5.2.2 ACE-inhibitory peptides

Angiotensin I-converting enzyme (ACE) removes two amino acids from the C-terminal angiotensin I to form the octapeptide angiotensin II which is the most hypertensive compound known. Angiotensin II reduces blood flow and thereby, decreases the renal excretion of the fluid and salts.

ACE-inhibitors from food proteins have been well known since 1979 when they were isolated from collagenase digests of gelatin for the first time. Furthermore numerous synthetic antihypertensive peptides have been synthesized corresponding to fragments of human caseins.

4.5.2.3 High F Ratio Oligopeptide

The peptide with high levels of branched chain amino acids (BCAA) and low levels of aromatic amino acids (AAA) is called high Fischer ratio oligopeptide. High Fischer ratio oligopeptide is an active peptide with a low molecular weight, derived from various food proteins.

Patients with severe hepatic disease generally have an amino acid imbalance characterized low levels of BCAA and high levels of AAA in their systemic blood. It has been reported that an increase in AAA levels in the brain leads to a decrease in the normal neurotransmitters and an increase in the neurologically inactive phenylethanolamine and octopamine, and that BCAA intake improves the plasma amino acid balance. Therefore, high Fischer ratio oligopeptides help to mitigate the symptoms of hepatic encephalopathy. Consequently, there is a market need for the development of an oligopeptide mixture with a high Fischer ratio.

Glossary

acid phosphatase	酸性磷酸酶
actin	肌动蛋白
adsorption	吸附作用
albumins	白蛋白
aleurone	谷物类种子的蛋白质微粒

FOOD CHEMISTRY

allantoinase	尿囊素酶
amylase	淀粉酶
angiotensin	血管紧张素
ammonia	氨
arginine	精氨酸
avidin	抗生素蛋白
calcium caseinate-phosphate	酪蛋白磷酸钙
carboxyl	羧基
casein	酪蛋白
chaotropic	离液序列高的
chelation	螯合作用
chromoprotein	色蛋白
chymosin	凝乳酶
citrate	柠檬盐类的
collagen	胶原
collagen	胶原蛋白
conalbumin	伴清蛋白
conjugated	共轭的
compressibility	可压缩性
cyanate	氰酸盐
cytochrome	细胞色素
deficient	缺乏的
denaturant	变性剂
denaturation	变性
derivative	衍生物
derived	衍生的
detergent	洗涤剂
dietary	饮食的
dilute acid	稀酸
dimmer	二聚物
disulfide bridge	二硫键
elastin	弹性蛋白
eletrophoretic	电泳的
flocculation	凝絮作用
foaming	起泡性
gelatin	胶质
gel	凝胶
gliadin	谷物醇溶蛋白

Globulins	球蛋白
glutamic	谷氨酸的
glutelins	谷蛋白
glutenin	麦谷蛋白
glycinin	大豆球蛋白
glycoprotein	糖蛋白
guanidine hydrochloride	盐酸胍
hemagglutination	红细胞凝集
histone	组织蛋白
hydrated	水合的
hydration	水和作用
hydrophilic	亲水性的
hydrophobic	疏水的
hydrostatic	静水压的
hydroxyl	羟基
isoelectric point	等电点
keratin	角蛋白
lactalbumin	乳白蛋白
lactoglobulin	乳球蛋白
lactoglobulin	乳球蛋白
legumelin	豆清蛋白
leucosin	麦清蛋白
lipoprotein	脂蛋白
lipovitellenin	卵黄脂磷蛋白
lipovitellin	卵黄磷蛋白
lysine	赖氨酸
lysine	赖氨酸
lysozyme	溶菌酵素
magnesium	镁
methionine	甲硫氨酸
mucoprotein	黏蛋白
myosin	肌球蛋白
net charge	净电荷
neurotransmitter	神经递质
nonreversible	不可逆的
nucleoprotein	核蛋白
oligopeptide	低聚肽,寡肽
oryzenin	米谷蛋白

ovalbumin	卵清蛋白
ovomucin	卵黏蛋白
ovomucoid	卵类黏蛋白
paracasein	副酪蛋白
peptone	蛋白胨
permittivity	介电常数
peroxidase	过氧化酶
phenylethanolamine	苯乙醇胺
phosphoprotein	磷蛋白,磷朊
phosphorylate	磷酸化
phosphorylation	磷酸化作用
polymorph	多晶体
polypeptide	多肽
precipitate	使……沉淀
prolamins	醇溶谷蛋白
proline	脯氨酸
protamiine	鱼精蛋白
proteolysis	蛋白酶解作用
proteolysis	水解作用
proteose	朊间质
proton	质子
rennet	凝乳
repulsion	斥力
rigid	非极性的
scleroproteins	硬蛋白
sedimentation rate	沉降系数
serine	丝氨酸
serum album	血清蛋白
shear	剪切力
spherical	球状的
subsonic pulse	次音波
sulfhydryl	巯基
sulfur	硫
surfactant	表面活性剂
tenacious	黏性强的
tendon	肌腱
tropocollagen	原胶原蛋白
trypsin	胰蛋白酶

tryptophan	色氨酸
ultracentrifuge	超高速离心机
vitellenin	脱脂卵黄脂磷蛋白
vitellin	脱脂卵黄磷蛋白
vitro-proteolysis	体外蛋白水解作用
whey	乳清
xanthine oxidase	黄嘌呤氧化酶
zein	玉米醇蛋白

FOOD CHEMISTRY

Chapter 5   Lipids

5.1   Introduction

5.1.1   Definition and Roles

Lipids consist of a broad group of compounds that are generally soluble in organic solvents including hexane, acetone, chloroform and petroleum ether, but not in water. Together with proteins and carbohydrates, they constitute the principal structural components of all living cells. Glycerol esters of fatty acids, which make up to 99% of the lipids of plant and animal origin, have been traditionally called fats and oils. The term oils and fats describe triacyglycerols (TAGs), formerly known as triglycerides, in liquid and solid state respectively. Everyday experience shows that oils and fats are interconvertible via melting as the surrounding temperature raised or lowered. An oil in the summer months may become a fat in winter, and a oil in the tropics may be a fat in temperate climates. Nevertheless owing to their generally high degree of saturation, animal derived TAGs tend to be fats whilst plant and fish derived TAGs are generally oils.

Food lipids are either consumed in the form of "visible" fats, which have been separated from original plant or animal sources, such as butter, lard, and culinary oil, or as constituents of basic foods, such as milk, cheese, and meat. The largest supply of vegetable oil comes from the seeds of soybean, cottonseed, peanut, and the oil-bearing trees of palm, coconut, and olive.

Lipids in food exhibit unique physical and chemical properties. Their composition, crystalline structure, melting properties, and ability to associate with water and other non-lipid molecules are especially important for their functional properties in many foods. During the processing, storage, and handling of foods, lipids undergo complex chemical changes and react with other food constituents, producing numerous compounds both desirable and undesirable to food quality.

Dietary lipids play an important role on nutrition. They supply calories and essential fatty acids, act as vitamin carriers, and increase the palatability of food, but they have been at the center of controversy for decades with respect to toxicity, obesity and disease.

5.1.2 Classification of Lipids

A classification of lipids proposed by Bloor (Table 5.1) is useful for us to distinguish many lipid substances.

Table 5.1 The classification of lipids

Simple lipids —Esters of fatty acids with alcohols	a. Fats: esters of fatty acids with glycerol b. Waxes: esters of fatty acids with long chain alcohols
Compound lipids —Compounds containing other groups in addition to an ester of a fatty acid with alcohol	a. Phospholipids (phosphatides): esters containing fatty acids, phosphoric acid, and other groups usually containing nitrogen b. Cerebrosides (glycolipids): compounds containing fatty acids, a carbohydrate and a nitrogen moiety, but no phosphoric acid c. Other compound lipids: sphingolipids and sulfolipids
Derived lipids —Substances derived from simple lipids or compound lipids and having general properties of lipids	a. Fatty acids b. Alcohols: usually long chain alcohols and sterols c. Hydrocarbons

Foods may contain any or all of these substances but those of greatest concern are the fats or glycerides and the phosphatides. The term "fats" is applicable to all triglycerides regardless of whether they are normally non-liquid or liquid at ambient temperatures. Liquid fats are commonly referred to as oils. Such oils as soybean oil, cottonseed oil and olive oil are of plant origin. Lard and tallow are examples of non-liquid fats from animals, yet the fat from horse is liquid at ambient temperatures and is referred to as horse oil.

Fats and oils also can be classed according to their "group characteristics". Five well-recognized groups are the milk fat group, the lauric acid group, the oleic-linoleic acid group, the linolenic acid group, and the animal depot-fats group.

5.1.3 Components of Triacyglycerols

Food oil (fat) is composed of triacyglycerol (95%-99%), it is obvious that the diversity of triacyglycerol comes from the diversity of fatty acids from Eq. 5.1.

$$\begin{array}{c} CH_2-OH \\ | \\ HO-C-OH \\ | \\ CH_2-OH \end{array} + 3R_iCOOH \longrightarrow \begin{array}{c} CH_2OCOR_1 \\ | \\ R_2OCOCH \\ | \\ CH_2OCOR_3 \end{array} \qquad (Eq.\ 5.1)$$

Even numbered, straight-chain saturated and unsaturated fatty acids make up the greatest proportion of the fatty acids of natural fats. However, it is now known that many other fatty acids may be present in small amounts. Some of these include odd carbon number fatty acids, branched-chain acids, and hydroxyl acids. These may occur in natural fats, as well as in processed fats. The latter category may, in addition, contain a variety of isomeric fatty acids which are not normally found in natural fats. It is customary to divide the fatty acids into different groups, for example, into saturated and unsaturated ones. This division is useful in food technology because saturated fatty acids have a much higher melting point than unsaturated ones, so the ratio of saturated fatty acids to unsaturated ones significantly affects the physical properties of fat or oil. Another common division is into short-chain, medium-chain, and long-chain fatty acids. Unfortunately, there is no generally accepted division of these groups. Generally, short-chain fatty acids have from 4 to 6 carbon atoms; medium-chain fatty acids, 8 to 12 carbon atoms; and long-chain fatty acids, 16 or more carbon atoms. However, some authors use the terms long- and short-chain fatty acid in a relative sense. In a fat containing fatty acids with 16 and 18 carbon atoms, the 16 carbon acid could be called the short-chain fatty acid. Yet another division differentiates between essential and nonessential fatty acids. The main fatty acids composing triacyglycerols in food oils are listed in Table 5.2.

Table 5.2 Main fatty acids in triacyglycerols

Abbreviation	Systematic name	Common name	Symbol
4:0	Butanoic	Butyric	B
6:0	Hexanoic	Caproic	H
8:0	Octanoic	Caprylic	Oc
10:0	Decanoic	Capric	D
12:0	Dodecanoic	Lauric	La
14:0	Tetradecanoic	Myristic	M
16:0	Hexadecanoic	Palmitic	P
16:1(n-7)	9-Hexadecenoic	Palmitoleic	Po
18:0	Octadecanoic	Stearic	St[a]
18:1(n-9)	9-Octadecenoic	Linoleic	
18:2(n-6)	9,12-Octadecadienoic	Linoleic	L
18:3(n-3)	9,12,15-Octadecatrienoic	Linolenic	Ln
20:0	Arachidic	Eicosanoic	Ad
20:4(n-6)	5,8,11,14-Eicosatetraenoic	Arachdonic	An
20:5(n-3)	5,8,11,14,17-Eicosapentaenoic	EPA	
22:1(n-9)	13-Docosenoic	Erucie	E
22:5(n-3)	7,10,13,16,19-Docosapentaenoic		
22:6(n-6)	4,7,10,13,16,19-Docosahexaenoic	DHA	

In earlier studies, major classes of triacylglycerols were separated on the basis of unsaturation (i.e., trisaturated, disaturated, diunsaturated, and triunsaturated) via fractional crystallization and oxidation-isolation methods. More recently, the techniques of stereospecific analysis made it possible for the detailed determinations of individual fatty acid distribution in each of the three positions of the triacylglycerols of many fats.

5.1.4　Edible Oils

Foods contain more or less fats (Table 5.3), and the population all over the world consume millions of tons oils every year (Table 5.4), most of which are from plant sources and few come from animals (fish).

Table 5.3　Fat content of some foods

Product	Fat(%)
Asparagus	0.25
Oats	4.4
Barley	1.9
Rice	1.4
Walnut	58
Coconut	34
Peanut	49
Soybean	17
Sundlower	28
Milk	3.5
Butter	80
Cheese	34
Hamburger	30
Beef cuts	10-30
Chicken	7
Ham	31
Cod	0.4
Haddock	0.1
Herring	12.5

Table 5.4　World consumption of vegetable and marine oils (USDA,1999)

Oil sources	Million metric tons
Soybean oil	24.5
Palm oil	21.2
Rapeseed (canola) oil	13.3
Sunflower seed oil	9.5
Peanut oil	4.3
Cottonseed oil	3.7
Coconut oil	3.2
Olive oil	2.4
Marine (fish) oil	1.2
Total	85.7

5.1.4.1 Plant Triglycerides

In general, plant seed oils contain more unsaturated fatty acids (Table 5.5) than animal fats, that's why most of them exist as liquid at room temperature. Common fatty acids show preferential placement of unsaturated fatty acids at the sn-2 position, i.e., the middle hydroxyl on the glycerol backbone. Linoleic acid is especially concentrated at this position. The saturated acids occur almost exclusively at positions 1, 3. In most cases, the individual saturated or unsaturated acids are distributed in approximately equal quantities between the sn-1 and the sn-3 positions.

The saturated fats of plant origin show a different distribution pattern. Approximately 80% of the triacylglycerols in cocoa butter are disaturated, with 18:1 concentrated in the 2 position and saturated fatty acids almost exclusively located in sn-1 and sn-3 position (Table 5.7).

Approximately 80% of the triacylglycerols in coconut oil are trisaturated, with lauric acid concentrated at the sn-2 positions, octanoic at the sn-3, and myristic and palmitic at the sn-1 positions.

Plants containing erucic acid, such as rapeseed oil, show considerable positional selectivity in placement of their fatty acids. Erucic acid is preferentially located at positions 1,3, but more of it is present at the sn-3 position than at the sn-1 position.

Table 5.5 Component fatty acids of some vegetable oils

Oil	Fatty Acid Wt%					
	16:0	18:0	18:1	18:2	18:3	Totla C18
Canola	4	2	56	26	10	96
Cottonseed	27	2	18	51	Trace	73
Peanut	13	2	38	31	Trace	83
Olive	10	2	78	7	—	90
Rice bran	16	2	42	37	1	84
Soybean	11	4	22	53	8	89
Sunflower	5	5	20	69	—	95
Sunflower high oleic	4	5	81	8	—	96
Palm	44	4	39	11	—	54
Cocoa butter	26	34	35	3	—	74

* Peanut also contains about 3% of 22:0 and 1% of 22:1.

5.1.4.2 Animal Triacylglycerols

Distribution patterns of fatty acids in triacylglycerols differ among animals and vary among parts of the same animal. Depot fat, the component fatty acids in it are shown in

Table 5.6, can be altered by changing dietary fat. In general, however, the saturated fatty acid content of the sn-2 position in animal fats is greater than that in plant fats, and the difference in composition between the sn-1 and sn-2 positions is also greater. In most animal fats the 16:0 acid is preferentially esterified at the sn-1 position and the 14:0 at the sn-2 position. Short chain acids in milk fat are selectively associated with the sn-3 position. The major triacylglycerols of beef fat are of the SUS type.

Pig fat is unique among animal fats. The 16:0 acid is significantly concentrated at the central position, the 18:0 acid is primarily located at the sn-1 position, 18:2 at the sn-3 position, and a large amount of oleic acid occurs at positions 3 and 1 (Table 5.7). The long polyunsaturated fatty acids, characteristic of marine oils, are preferentially located at the sn-2 position.

Table 5.6 Component fatty acids of animal depot fats

Animal	Fatty Acids Wt%						
	14:00	16:0	16:1	18:0	18:1	18:2	18:3
Pig	1	24	3	13	41	10	1
Beef	4	25	5	19	36	4	Trace
Sheep	3	21	2	25	34	5	3
Chicken	1	24	6	6	40	17	1
Turkey	1	20	6	6	38	25	2

Table 5.7 Positional distribution of fatty acid in pig fat and cocoa butter

Fat	Position	Fatty Acid (Mole %)					
		14:0	16:0	16:1	18:0	18:1	18:2
Pig fat	1	0.9	9.5	2.4	29.5	51.3	6.4
	2	4.1	72.3	4.8	2.1	13.4	3.3
	3	0	0.4	1.5	7.4	72.7	18.2
Cocoa butter	1	—	34.0	0.6	50.4	12.3	1.3
	2	—	1.7	0.2	2.1	87.4	8.6
	3	—	36.5	0.3	52.8	8.6	0.4

5.2 Physical Properties

5.2.1 Melting Point and Polymorphism

The melting point is the temperature at which a kind of lipid transforms from the solid to the liquid state. The melting temperature for fats is generally broad since fats are a blend of different TAGs (triacyglycerols).

The melting point of a kind of fat is determined by the type of fatty acid constituents and their position on the glycerol backbone. Variables that affect the melting point of pure fatty acids probably affect triacylglycerols in a similar way. The following discussion applies to fatty acids and TAGs interchangeably. The melting point of a fatty acid increases with the chain length and with increasing degrees of saturation. Geometric isomerism also affects the melting point, with trans fatty acids possessing higher melting points compared to their cis isomers. The order of increasing melting point (saturated fatty acid > trans fatty acids > cis fatty acids) also showed in Tables 5.8 and 5.9, can be understood based on the model for the melting process. Crystallization requires the presence of multiple nuclei within the oil phase. Each nuclei then grows in size as more molecules adhere to the crystal surface. The rate of crystallization growth is described by the first order kinetics. Over the temperature range 0.5-17 ℃ the activation energy for fat crystallization is negative (-38.5 kJ/mole), which implies that the rate of crystallization increases with decreasing temperature.

During melting the ordered crystalline array of TAG molecules is transformed into the disordered and mobile liquid state. This requires an input of kinetic energy in the form of heat. The generally low melting point of cis fatty acids is explained by the marked "kink" within their structures. This irregular shape leads to the inefficient packing of cis fatty acids as compared with trans fatty acids and saturated fatty acids within a crystalline network.

Table 5.8 Melting points of some saturated fatty acids

Abbreviated designation	Structure	Systematic name	Common name	Melting point (℃)
A. Even numbered straight chain fatty acids				
4:0	$CH_3(CH_2)_2COOH$	Butanoic acid	Butyric acid	-7.9
6:0	$CH_3(CH_2)_4COOH$	Hexanoic acid	Caproic acid	-3.9
8:0	$CH_3(CH_2)_6COOH$	Octanoic acid	Caprylic acid	16.3
10:0	$CH_3(CH_2)_8COOH$	Decanoic acid	Capric acid	31.3
12:0	$CH_3(CH_2)_{10}COOH$	Dodecanoic acid	Lauric acid	44.0
14:0	$CH_3(CH_2)_{12}COOH$	Tetradecanoic acid	Myristic acid	54.4
16:0	$CH_3(CH_2)_{14}COOH$	Hexadecanoic acid	Palmitic acid	62.9
18:0	$CH_3(CH_2)_{16}COOH$	Octadecanoic acid	Stearic acid	69.6
20:0	$CH_3(CH_2)_{18}COOH$	Eicosanoic acid	Arachidic acid	75.4
22:0	$CH_3(CH_2)_{20}COOH$	Docosanoic acid	Behenic acid	80.0
24:0	$CH_3(CH_2)_{22}COOH$	Tetracosanoic acid	Lignoceric acid	84.2
26:0	$CH_3(CH_2)_{24}COOH$	Hexacosanoic acid	Cerotic acid	87.7

(To be continued)

(Table 5.8)

Abbreviated designation	Structure	Systematic name	Common name	Melting point (℃)
B. Odd numbered straight chain fatty acids				
5:0	$CH_3(CH_2)_3COOH$	Pentanoic acid	Valeric acid	−34.5
7:0	$CH_3(CH_2)_5COOH$	Heptanoic acid	Enanthic acid	−7.5
9:0	$CH_3(CH_2)_7COOH$	Nonanoic acid	Pelargonic acid	12.4
15:0	$CH_3(CH_2)_{12}COOH$	Pentadecanoic acid		52.1
17:0	$CH_3(CH_2)_{15}COOH$	Heptadecanoic acid	Margaric acid	61.3
C. Branched chain fatty acids				
	(structure) COOH	2, 6, 10, 14-Tetramethy-pentadecanoic acid	Pristanic acid	
	(structure) COOH	3, 7, 11, 15-Tetramethyl-hexadecanoic acid	Phytanic acid	

Table 5.9 Melting points of some unsaturated Fatty Acids

Abbreviated designation	Structure	Common name	Melting point (℃)
A. Fatty acids with nonconjugated double bonds			
	ω9-Family		
18:1(9)	$CH_3-(CH_2)_n-CH=CH-CH_2-(CH_2)_6-COOH$	Oleic acid	13.4
22:1(13)	$-(CH_2)_{10}-COOH$	Erucic acid	34.7
24:1(15)	$-(CH_2)_{12}-COOH$	Nervonic acid	42.5
	ω6-Family		
18:2(9,12)	$CH_3-(CH_2)_4-(CH=CH-CH_2)_2-(CH_2)_6-COOH$	Linoleic acid	−5.0
18:3(6,9,12)	$-(CH=CH-CH_2)_3-(CH_2)_3-COOH$	γ-Linolenic acid	
20:4(5,8,11,14)	$-(CH=CH-CH_2)_4-(CH_2)_2-COOH$	Arachidonic acid	−49.5
	ω3-Family		
18:3(9,12,15)	$CH_3-CH_2-(CH=CH-CH_2)_3-(CH_2)_6-COOH$	α-Linolenic acid	−11.0
20:5(5,8,11,14,17)	$-(CH=CH-CH_2)_5-(CH_2)_2-COOH$	EPA[a]	
22:6(4,7,10,13,16,19)	$-(CH=CH-CH_2)_6-CH_2-COOH$	DHA[a]	
B. Fatty acids with nonconjugated trans-double bonds			
18:1(tr9)	$CH_3-(CH_2)_n-CH^{tr}=CH-(CH_2)_7-COOH$	Elaidic acid	46
18:2(tr9, tr12)	$CH_3-(CH_2)_4-CH^{tr}=CH-CH_2-CH^{tr}=CH-(CH_2)_n-COON$	Linolelaidic acid	28
C. Fatty acids with conjugated double bonds			
18:2(9, tr11)	$CH_3-(CH_2)_5-CH^{tr}=CH-CH^{tr}=CH-(CH_2)_7-COOH$		
18:3(9, tr11, tr13)	$CH_3-(CH_2)_3-CH=CH-CH^{tr}=CH-CH^{tr}=CH-(CH_2)_n-COOH$	α-Eleostearic acid	48
18:3(tr9, tr11, tr13)	$CH_3-(CH_2)_n-CH^{tr}=CH-CH^{tr}=CH-CH^{tr}=CH-(CH_2)_n-COOH$	β-Eleostearic acid	71.5
18:4(9,11,13,15)[b]	$CH_3-CH_2-(CH=CH)_4-CH_2-(CH_2)_n-COOH$	Parinaric acid	85

[a] EPA: Eicosapentanoic acid, DHA: Docosahexanoic acid.

[b] Geometry of the double bond was not determined.

Polymorphism refers to the existence of more than one crystal form in fat, and it results from different patterns of molecular packing in fat crystals. Fat may occur in three basic polymorphs designated α (alpha), β' (beta prime), and β (beta). The α form is the least stable and has the lowest melting point; the β form is the most stable and has the highest melting point; the β' form is the intermediate in stability and the melting point. Polymorphic transformations occur from α to β' to β and are irreversible (Fig. 5.1).

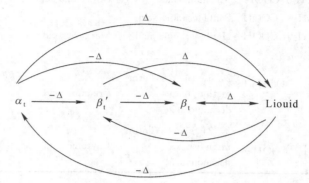

Fig. 5.1 Polymorphic transitions between different crystal forms of triacylglycerols

When a fat is cooled rapidly the α polymorph is produced, which is usually converted to the β' form quickly. In commerciall fat produces such as shortenings and margarines, β' is the desirable form because β' crystals are small and result in smooth texture and good functionalities. Depending on fatty acid and glyceride composition, fats may remain stable in the β' form or may convert to the β form. The latter will result in a large increase in crystal size and grainy texture. Different ratios of β' and β crystals can exist in fats. The melting point of α, β' and β forms of some simple triacylglycerols are given in Table 5.10.

The smallest repeating unit in a crystal is known as a subcell. Fig. 5.2 shows the crystal arrangements of the three polymorphic forms.

Table 5.10 Triacylglycerols and their polymorphic forms

Compound	Melting point(℃) of polymorphic form		
	α	β'	β
Tristearin	55	63.2	73.5
Tripalmitin	44.7	56.6	66.4
Trimyristin	32.8	45.0	58.5
Trilaurin	15.2	34	46.5
Trilaurin	−32	−12	4.5-5.7
1,2-Dipalmitoolein	18.5	29.8	34.8
1,3-Dipalmitoolein	20.8	33	37.3

(To be continued)

(Table 5.10)

Compound	Melting point(℃) of polymorphic form		
	α	β′	β
1-Palmito-3-stearo-2-olein	18.2	33	39
1-Palmito-2-stearo-3-olein	26.3	40.2	
2-Palmito-1-stearo-3-olein	25.3	40.2	
1,2-Diacetopalmitin	20.5	21.6	42.3

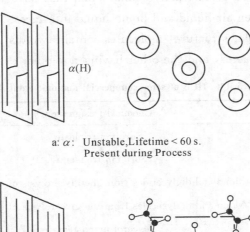

a: α: Unstable, Lifetime < 60 s.
Present during Process

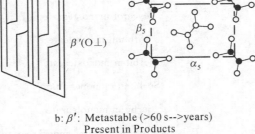

b: β′: Metastable (>60 s --> years)
Present in Products

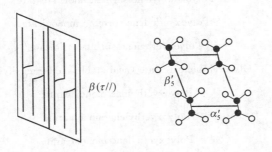

c: β: Stable

Fig. 5.2 Representation of the packing of triacylglycerols in the three main polymorphic forms

5.2.2 Emulsions and Emulsifiers

An emulsion is a heterogeneous system, consisting of one immiscible liquid intimately dispersed in another one, in the form of droplets with a diameter generally over 0.1 μm. These systems have a minimal stability, which can be enhanced by surface-active agents and some other substances. In many oil-contained foods, oils usually exist in emulsions. Emulsions usually contain two phases, oil and water. If water is the continuous phase and oil is the disperse phase, the emulsion is of the oil in water (O/W) type. In the reverse case, the emulsion is of the water in oil (W/O) type. A third material or combination of several materials is required to confer stability upon the emulsion. These are surface-active agents called emulsifiers. The action of emulsifiers can be enhanced by the presence of stabilizers. Emulsifiers are surface-active compounds that have the ability to reduce the interfacial tension between air-liquid and liquid-liquid interfaces. This ability is the result of an emulsifier's molecular structure: molecules contain two distinct sections, one having polar or hydrophilic character, and the other having non-polar or hydrophobic properties.

Table 5.11 HLB of some commercial nonionic emulsifiers

Trade name	Chemical Designation	HLB
Span 85	Sorbitan trioleate	1.8
Span 65	Sorbitan tristearate	2.1
Atmos 150	Mono-and diglycerides from the glycerolysis of edible fats	3.2
Atmul 500	Mono-and diglycerides from the glycerolysis of edible fats	3.5
Atmul 84	Glycerol monostearate	3.8
Span 80	Sorbitan monooleate	4.3
Span 60	Sorbitan monostearate	4.7
Span 40	Sorbitan monopalmitate	6.7
Span 20	Sorbitan monolaurate	8.6
Tween 61	Polyoxyethylene sorbitan monostearate	9.6
Tween 81	Polyoxyethylene sorbitan monooleate	10.0
Tween 85	Polyoxyethylene sorbitan trioleate	11.0
Arlacel 165	Glycerol monostearate (acid stable, self-emulsifying)	11.0
Myrj 45	Polyoxyethylene monostearate	11.1
Atlas G-2127	Polyoxyethylene monolaurate	12.8
Myrj 49	Polyoxyethylene monostearate	15.0
Myrj 51	Polyoxyethylene monostearate	16.0

Source: From WC. Griffin, Emulsions, in Kirk-Othmer Encyclopedia of Chemical Technology, 2nd ed., Vol. 8, pp. 117-154, 1965, John Wiley & Sons.

The behavior of an emulsifier in emulsification is mostly determined by the relative size of the hydrophilic and hydrophobic sections in the emulsifier molecule. To make the proper selection of the emulsifier for a given application, the so-called HLB (hydrophile-lipophile balance) system was developed. It is a numerical expression for the relative simultaneous attraction of an emulsifier for water and for oil. The HLB of an emulsifier is an indication of how it will behave but not how efficient it is. Emulsifiers with low HLB tend to form W/O emulsions, those with intermediate HLB form O/W emulsions, and those with high HLB are solubilizing agents. The HLB scale of nonionic emulsifiers goes from 0 to 20, in theory at least, since at each end of the scale the compounds would have little emulsifying activity. The HLB values of some commercial nonionic emulsifiers are given in Table 5.11.

Foods contain many natural emulsifiers, of which phospholipids are the most common. Crude phospholipids mixtures obtained by degumming of soya oil are utilized extensively as food emulsifiers and are known as soya-lecithin. This product contains a variety of phospholipids, not just lecithin.

Emulsifiers can be tailor-made to serve in many food emulsion systems. Probably the most widely used is the group of monoglycerides obtained by glycerolysis of fats. Reaction of an excess of glycerol with a fat under vacuum at high temperature and in the presence of a catalyst results in the formation of so called monoglycerides. These are mixtures of mono-, di-, and triglycerides. Only the l-monoglycerides are active as emulsifiers. By molecular distillation under the high vacuum, a product can be obtained in which the l-monoglyceride content exceeds 90 percent.

Emulsions are stabilized by a variety of compounds, mostly macromolecules such as proteins and starches.

Emulsifiers have many additional functions in foods. They form complexes with food components, resulting in modified physical properties of the food system. For example, amylose-complexing effect of emulsifiers. This effect is useful for improving the shelf life of bread (anti-firming effect) and modifying the physical characteristics of rice and wheat products, such as rice cake, pasta, noodle, and similar foods.

5.3 Chemical Reactivity of Fats and Oils

5.3.1 Oxidation

Lipid oxidation is one of the major causes of food spoilage. It is of great economic concern to the food industry because it leads to the development in edible oils and fat-containing foods, of various off flavors and off odors generally called rancid (oxidative rancidity), which render these foods less acceptable. In addition, oxidative reactions can decrease the nutritional quality of food, and certain oxidation products are potentially

FOOD CHEMISTRY

toxic. On the other hand, under certain conditions, a limited degree of lipid oxidation is sometimes desirable, as in aged cheeses and some fried foods.

Fat oxidation occurs when activated oxygen species react with unsaturated fatty acids. The process is a free radical chain reaction. The key intermediates are hydroperoxides, which degrade to volatile aldehydes and ketones with strong off-flavors.

5.3.1.1 Autoxidation

It is generally agreed that "autoxidation", the reaction with molecular oxygen via a self-catalytic mechanism, is the main reaction involved in oxidative deterioration of lipids. Although photochemical reactions have been known for a long time, only recently the role of photosensitized oxidation and its interaction with autoxidation have emerged. In foods, the lipids can be oxidized by both enzymatic and non-enzymatic mechanisms.

The unsaturated bonds that present in all fats and oils represent active centers that, among other things, may react with oxygen. This reaction leads to the formation of primary, secondary, and tertiary oxidation products that may make the fat or fat-containing foods unsuitable for consumption.

The process of autoxidation and the resoulting deterioration in flavor of fats and fatty foods are often described by the term "rancidity". Usually rancidity refers to oxidative deterioration but, in the field of dairy science, rancidity refers usually to hydrolytic changes resulting from the enzyme activity. Flavor reversion is the term used for the objectionable flavors that develop in oils containing linolenic acid. This type of oxidation is produced with considerably less oxygen than common oxidation. A type of oxidation similar to reversion may take place in dairy products, where a very small amount of oxygen may result in intense oxidation off-favors. It is interesting to note that the linolenic acid content of milk fat is quite low.

The autoxidation reaction can be divided into the following three stages: initiation, propagation, and termination. In the initiation stage, hydrogen is abstracted from an olefinic compound to yield a free radical.

$$RH \longrightarrow R\cdot + H\cdot$$

The removal of hydrogen takes place at the carbon atom next to the double bond and can be brought about by the action of, for instance, light or metals. Once a free radical has been formed, it will combine with oxygen to form a peroxy-free radical, which can in turn abstract hydrogen from another unsaturated molecule to yield a peroxide and a new free radical, thus starting the propagation reaction. This reaction may be repeated up to several thousand times and has the nature of a chain reaction.

$$R\cdot + O_2 \longrightarrow ROO\cdot$$
$$ROO\cdot + RH \longrightarrow ROOH + R\cdot$$

The propagation can be followed by termination if the free radicals react with themselves to yield non-active products, as shown here:

$$R\cdot + R\cdot \longrightarrow R\text{-}R$$

$$R\cdot + ROO\cdot \longrightarrow ROOR$$
$$ROO\cdot + ROO\cdot \longrightarrow ROOR + O_2$$

The hydroperoxides formed in the propagation stage of the reaction are the primary oxidation products. The hydroperoxides are generally unstable and decompose into the secondary oxidation products, which include a variety of compounds, including carbonyls, which are the most important. The peroxides have no importance to the flavor deterioration, which is wholly caused by the secondary oxidation products. In the initial stages of the reaction, the amount of hydroperoxides increases slowly; this stage is termed as the induction period. At the end of the induction period, there is a sudden increase in peroxides content. Because peroxides are easily determined in fats, the peroxide value is frequently used to measure the progress of oxidation. Organoleptic changes are more closely related to the secondary oxidation products, which can be measured by various procedures, including the benzidine value, which is related to aldehyde decomposition products. As the aldehydes are themselves oxidized, fatty acids are formed; these free fatty acids may be considered tertiary oxidation products. The length of the induction period, therefore, depends on the method used to determine oxidation products.

Among the many factors that affect the rate of oxidation are the following:
- Degree of unsaturation of lipids
- Amount of oxygen present
- Presence of antioxidants
- Presence of prooxidants, especially copper, and some organic compounds such as heme-containing molecules and lipoxidase
- Nature of packaging material
- Light exposure
- Temperature of storage

Although even saturated fatty acids may be oxidized, the rate of oxidation greatly depends on the degree of unsaturation. In the series of 18 carbon-atom fatty acids 18:0, 18:1, 18:2, 18:3, the relative rate of oxidation has been reported to be in the ratio of 1:100:1200:2500(Table 5.12)

Table 5.12 Induction period and relative rate of oxidation for fatty acids at 25 ℃

Fatty acid	Number of allyl groups	Induction period (h)	Oxidation rate(relative)
18:0	0		1
18:1(9)	1	82	100
18:2(9,12)	2	19	1,200
18:3(9,12,15)	3	1.34	2,500

As Fig. 5.3 indicates, peroxides are labile and may be transformed into secondary oxidation products.

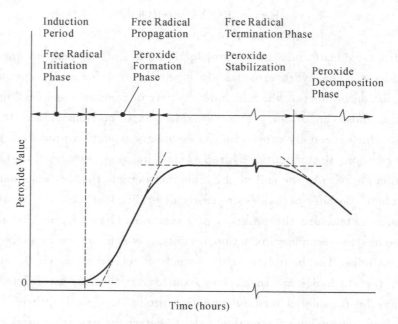

Fig. 5.3 Peroxide formation and decomposition as a function of time

Removal of oxygen from foods will prevent oxidation, however, in practice, this is not easy to accomplish in many cases.

At high temperatures (100 to 140 ℃) such as those used in the accelarated tests for oil stability (active oxygen method), formic acid is produced, which is the main reaction product resulted from aldehyde decomposition, and other short-chain acids from acetic to caproic were also formed.

Trace metals, especially copper and iron, will catalyze fat oxidation, thus, metal deactivators such as citric acid can be used to reduce the adverse effect of metals. Lipoxygenase (lipoxidase) and heme compounds act as catalysts of lipid oxidation.

Antioxidants can be very effective in slowing down oxidation and increasing the induction period. Many foods contain natural antioxidants, and tocopherols are the most important one. They are present in greater amounts in vegetable oils than in animal fats, which may explain the former's greater stability. Synthetic antioxidants may be added as in foods.

Antioxidants act by reacting with free radicals, thus terminating the reaction chain. The antioxidant AH may react with the fatty acid free radical or with the peroxy free radical to form $A\cdot$.

$$AH + R\cdot \longrightarrow RH + A\cdot$$
$$AH + RO_2\cdot \longrightarrow RO_2H + A\cdot$$

The antioxidant free radical deactivated by the further oxidation to quinones, thus terminating the chain. Only phenolic compounds that can easily produce quinones are active as antioxidants. At high concentration antioxidants may have a pro-oxidant effect and one of the reactions may be as follows:

$$A^\cdot + RH \longrightarrow AH + R^\cdot$$

The highly active antioxidants that are used in the food industry are active at about 10 to 50 parts per million (mg/kg). Chemical structure of the antioxidants is the most important factor affecting their activity. The number of synthetic antioxidants permitted in foods is limited, and the structure of the most widely used compounds is shown in Fig. 5.4.

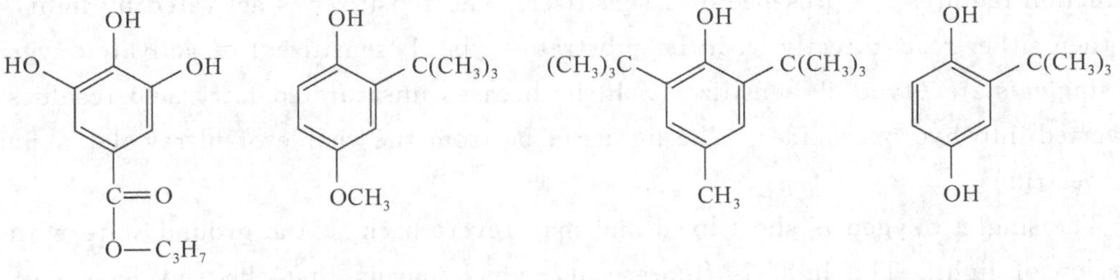

Fig. 5.4 Structure of propyl gallate (PG), butylated hydroxyanisole (BHA), butylated hydroxy toluene (BHT) and tert-butyl hydroquinine (TBHQ)

Propyl gallate (PG) is more soluble in water than in fats. Butylated hydroxyanisole (BHA) is heat resistant and nonvolatile with steam, making it useful for frying oils and in baked products, is considered to have carry-through propereies, but butylated hydroxy toluene (BHT) does not, because it is volatile with steam. The compound tert-butyl hydroquinine (TBHQ) is used for its effectiveness in increasing oxidative stability in polyunsaturated oils and fats. It also provides carry-through protection for fried foods.

Antioxidants are frequently used in combination or together with synergists. The latter are frequently metal deactivators that have the ability to chelate metal ions. It has been pointed out that fatty acid composition can explain only about half of the oxidative stability of a vegetable oil, and the other half can be contributed to minor components including tocopherols, metals, pigments, free fatty acids, phenols, phospholipids, and sterols.

Some physical process for food may affect oxidation, for example, finely divided foods have a higher surface area for gaseous exchange and oxidation. grinding and shearing can cause the release of oxidation catalysts. Refrigeration is not an efficient method for preventing oxidation because the solubility of oxygen increases at low temperatures. Thus physical methods for preventing oxidation include the use of packaging, and packaging materials have been developed that contain active (reducing) agents. Another important innovation is vacuum packaging whereby the food container is evacuated to remove oxygen. Vacuum packaging is especially effective with meat and fried products.

5.3.1.2 Photo-oxidation

Oxidation of lipids, in addition to the free radical process, can be brought about by at least two other mechanisms—photo-oxidation and enzymatic oxidation by the lipoxygenase. The light-induced oxidation or photo-oxidation results from the reactivity of an excited state of oxygen, known as the singlet oxygen. Ground-state or normal oxygen is the triplet oxygen. The activation energy for the reaction of normal oxygen with an unsaturated fatty acid is very high, of the order of 146 to 273 kJ/mol. When oxygen is converted from the ground state to the singlet state, energy is taken up amounting to 92 kJ/mol, and in this state the oxygen is much more reactive. The singlet-state oxygen production requires the presence of a sensitizer. The sensitizer is activated by light, and can then either react directly with the substrate (type I sensitizer) or activate oxygen to the singlet state (type II sensitizer). In both cases unsaturated fatty acid residues are converted into hydroperoxides. The light can be from the visible or ultraviolet region of the spectrum.

The singlet oxygen is short-lived and may revert back to the ground state with the emission of light. The light is fluorescent, which means that the wavelength of the emitted light is higher than that of the light which is absorbed for the excitation. The reactivity of the singlet oxygen is 1,500 greater than that of the ground-state oxygen. Compounds that can act as sensitizers are widely occurring in food components, including chlorophyll, myoglobin, riboflavin, and heavy metals. Most of these compounds promote type II oxidation reactions. In these reactions, the sensitizer is transformed into the activated state by light. The activated sensitizer then reacts with oxygen to produce the singlet oxygen. The singlet oxygen can react directly with unsaturated fatty acids. The singlet oxygen reacts directly with the double bond by addition, and shifts the double bond one carbon away. The singlet oxygen attacks on the linoleate to produce four hydroperoxides as shown in Fig. 5.5. Photo-oxidation has no induction period, but the reaction can be quenched by carotenoids that effectively compete for the singlet oxygen and bring it back to the ground state.

Phenolic antioxidants do not protect fats from the oxidation by the singlet oxidation. However, the antioxidant ascorbyl palmitate is an effective singlet oxygen quencher. Carotenoids are widely used as quenchers.

The combination of light and sensitizers is present in many foods packed in transparent containers in bright supermarkets. The light-induced deterioration of milk has been studied extensively. Sattar *et al.* (1976) reported on the light-induced flavor deterioration of several oils and fats, of the five fats examined, milk fat and soybean oil were most susceptible and corn oil was least susceptible to the singlet oxygen attack. The effect of temperature on the rate of oxidation of illuminated corn oil was reported by Chahine and deMan (1971). They found that temperature has an important effect on photo-oxidation rates, but even freezing does not completely prevent the oxidation.

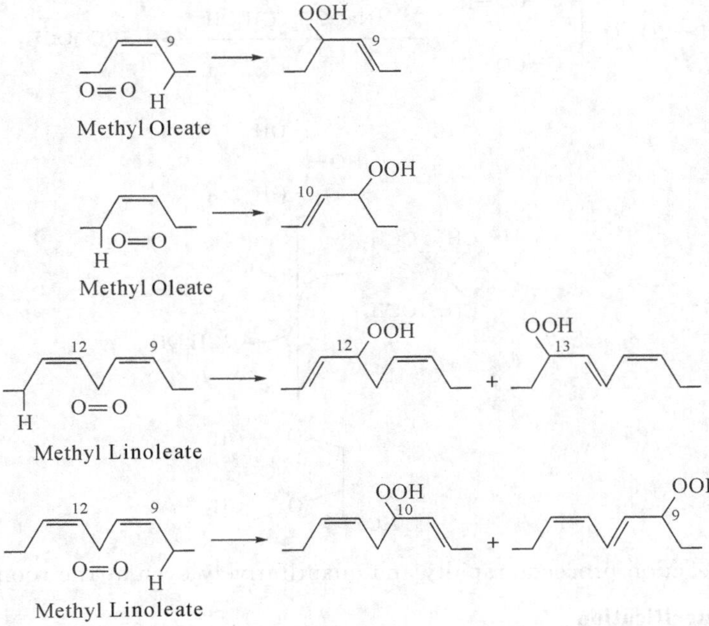

Fig. 5.5 Photo-oxidation. Singlet oxygen attack on oleate produces two hydroperoxides; linoleate yields four hydroperoxides

5.3.2 Other Reaction

Hydrolysis, methanolysis and interesterification are the most important chemical reactions for tryacylglycerols (TGs).

5.3.2.1 Hydrolysis

The fat or oil is cleaved or saponified by the treatment with alkali (e. g. alcoholic KOH):

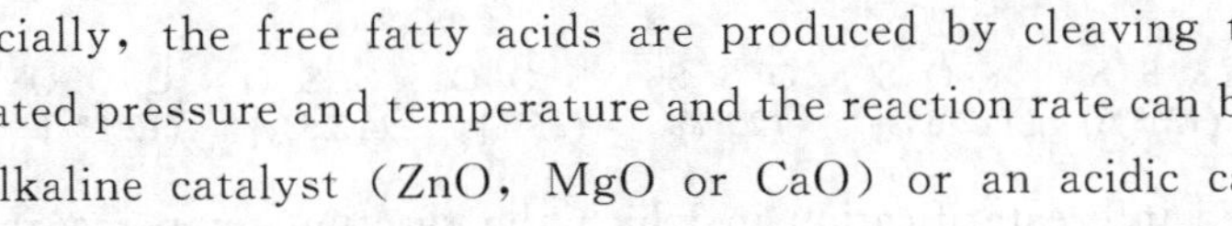

After saponification, the free fatty acids are recovered as alkali salts (commonly called soaps). Commercially, the free fatty acids are produced by cleaving triglycerides with steam under elevated pressure and temperature and the reaction rate can be increased in the presence of an alkaline catalyst (ZnO, MgO or CaO) or an acidic catalyst (aromatic sulfonic acid) and then acidification.

5.3.2.2 Methanolysis

The fatty acids in TGs are usually analyzed by gas liquid chromatography, not as free acids, but as methyl esters, because the later has lower boiling pornt, and can be gasified easily. The required trans-esterification is most often achieved by Na-methylate (sodium methoxide) in methanol and in the presence of 2,2-dimethoxypropane to bind the released glycerol:

$$\text{R-CO-O}\begin{bmatrix}\text{O-CO-R}\\ \text{O-CO-R}\end{bmatrix} \xrightarrow{\text{NaOCH}_3/\text{CH}_3\text{OH}} 3\text{RCOOCH}_3$$

(with glycerol HO-CH-OH / HO- released, and acetone dimethyl acetal $(CH_3)_2C(OCH_3)_2$ reacting to form the isopropylidene glycerol acetal + $2CH_3OH$)

Thus, the reaction proceeds rapidly and quantitatively even at the room temperature.

5.3.2.3 Interesterification

This reaction is of industrial importance since it can change the physical properties of fats or oils or their mixtures without altering chemical structure of the fatty acids. Both intra- and inter-molecular acyl residue exchange occur in the reaction until an equilibrium is reached which depends on the structure and composition of the TG molecules. The usual catalyst for the interesterification is Na-methylate or Na-ethylate.

The principle of the reaction will be elucidated by using a mixture of tristearin (SSS) and triolein (OOO) or stearodiolein (OSO). Two types of interesterification are recognized:

ⅰ) A single-phase interesterification where the acyl residues are randomly distributed:

$$\begin{array}{c}\text{SSS} + \text{OOO}\\ (50\%) \quad (50\%)\\ \downarrow (\text{NaOCH}_3)\\ \text{SSS} \quad \text{SOS} \quad \text{OSS} \quad \text{SOO} \quad \text{OSO} \quad \text{OOO}\\ (12.5\%) \ (12.5\%) \ (25\%) \ (25\%) \ (12.5\%) \ (12.5\%)\end{array}$$

ⅱ) A directed interesterification in which the reaction temperature is lowered until the higher melting and least soluble TG molecules in the mixture crystallize. These molecules cease to participate in further reactions, thus the equilibrium is continuously changed. Hence, a fat (oil) can be divided into high and low melting point fractions, e.g:

$$\underset{\substack{S\ S\ S\\(33.3\%)}}{} \quad \underset{\substack{O\ O\ O\\(66.7\%)}}{} \xleftarrow{(NaOCH_3)} O\ S\ O$$

The interesterification can also be catalyzed by the enzyme. For example the reaction between tripalmitin (PPP) and oleic acid or its non glycerol esters such as methyl or ethyl oleate needs to be catalyzed by 1,3-specific lipase to produce 1,3-Dioleoyl-2-palmitoyl glycerol (OPO), which is the characteristic triacylglycerol of the human milk fat.

5.3.2.4 Iodine Binding

The number of double bonds present in oil or fat can be determined through their iodine numbers. The fat or oil is treated with a halogen reagent which reacts only with double bonds. Substitution reactions generating hydrogen halides must be avoided. IBr in an inert solvent, such as glacial acetic acid, is a suitable reagent:

$$\text{>C=C<} + IBr \longrightarrow \text{>C}^{\oplus}\text{-C<}\ Br^{\ominus} \longrightarrow -\overset{I}{\underset{|}{C}}-\overset{Br}{\underset{|}{C}}-$$

The number of double bonds is calculated by titrating the unreacted IBr reagent with thiosulfate.

5.3.2.5 Hydrogenation

In the presence of a suitable catalyst, e.g. Ni, hydrogen can be added to the double bond of the unsaturated fatty acids. This heterogeneous catalytic hydrogenation occurs stereo selectively as a cis-addition. Catalyst-induced isomerization from an isolene-type fatty acid to a conjugated fatty acid occurs with fatty acids with several double bonds:

(1) Isomerization
(2) Hydrogenation

Since diene fatty acids form a more stable complex with a catalyst than do monoene fatty acids, the former are preferentially hydrogenated. Since the nature is not an abundant source of the solid fats which are required in food processing, the partial and selective hydrogenation, just referred to, plays an important role in the industrial processing of fats and oils. After hydrogenation, the fat or oil will become more stable and the melting point will increase, which make the fat more suitable for frying and shortening, but the tran-fatty acids generated during the hydrogenation may cause health concerns.

5.4 Functional Lipids

5.4.1 Functional Fatty Acids

5.4.1.1 Medium Chain Fatty Acids

Medium-chain fatty acids (MCFA) are fatty acids with aliphatic tails of 8-12 carbons, which can form medium-chain triglycerides. Medium chain triglycerides (MCT) are unique non-carbohydrate high and rapidly available energy for malabsorption syndrome cases and the infant care. Structure lipids with a MCT backbone and essential fatty acid or other polyunsaturated fatty acids built into the triglyceride molecule are best for patients, particularly for the critically ill.

Besides the nutritive value, MCFA have been shown to have significant antibacterial and antifungal activity, which together with the consideration of their natural properties allows them to be utilized in the food industry as food preservative. Actually, when used as this goal, monoesters of MCFA are preferred rather than the relevant fatty acids because the former usually has stronger antimicrobial activity and at the same time the monoglycerides of MCFA can act as food emulsifier. Among the monoesters of MCFAs, glycerol monolaurte (GML) is the best choice at most cases due to the stronger antimicrobial activity and better flavor property of GML than other monoesters of MCFAs.

5.4.1.2 Polyunsaturated Fatty Acids

Polyunsaturated fatty acids (PUFA) are fatty acids that contain more than one double bond in their backbone. This class includes many important compounds, such as essential fatty acids (linoleic acid and linolinic acid) and those that give oils their characteristic properties. The relationship of PUFA with some widely occurred human diseases such as cardiovascular and cerebrovascular is known gradually in the past decades.

Methylene-Interrupted Polyenes

These fatty acids have 2 or more cis double bonds that are separated from each other by a single methylene group. The essential fatty acids are all omega-3 and -6 methylene-interrupted fatty acids. Examples of omega-3 fatty acids are alpha-linolenic acid (ALA), eicosapentaenoic acid (EPA), docosahexaenoic acid (DHA), and the typical omega-6 fatty acids are linoleic acid (LA) and gama-linolenic acid (GLA).

Conjugated Fatty Acids

Conjugated fatty acids are polyunsaturated fatty acids in which at least one pair of double bonds are separated by only one single bond. An example of conjugated fatty acids is conjugated linoleic acids (CLA), which are a family of at least 28 isomers of linoleic acid found rarely in the meat and dairy products derived from ruminants. As the name implies, the double bonds of CLAs are conjugated, with only one single bond between them.

5.4.2 Phytosterol

Phytosterols (also called plant sterols) are a group of steroid alcohols naturally occurring in plants. Phytosterols including stigmasterol, β-sitosterol, campesterol, and brassicasterol occur naturally in small quantities in vegetable oils. The structure of the plant sterols showed in Fig. 5.6, and the sterol content of common edible oils listed in Table 5.13. Phytosterols are white powders with mild, characteristic odor, insoluble in water and soluble in alcohols. They have applications in medicines and cosmetics and as a food additive taken to lower cholesterols.

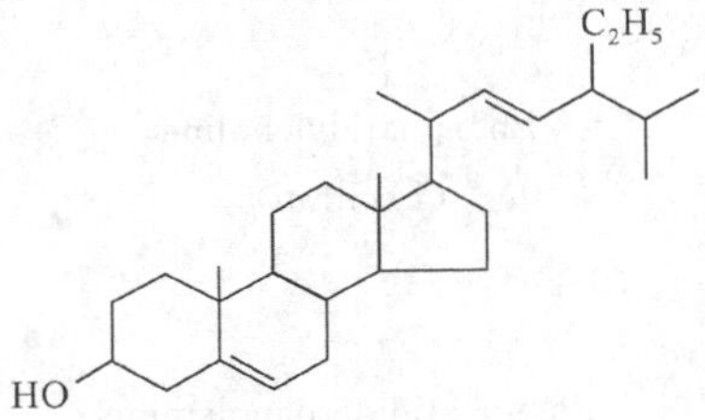

Stigmasterol

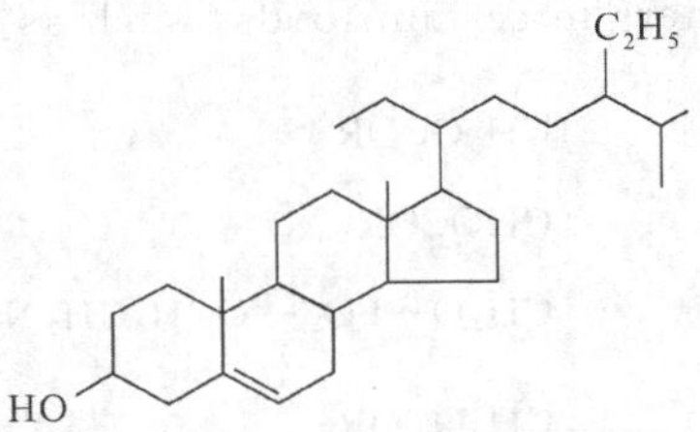

β-Sitosterol

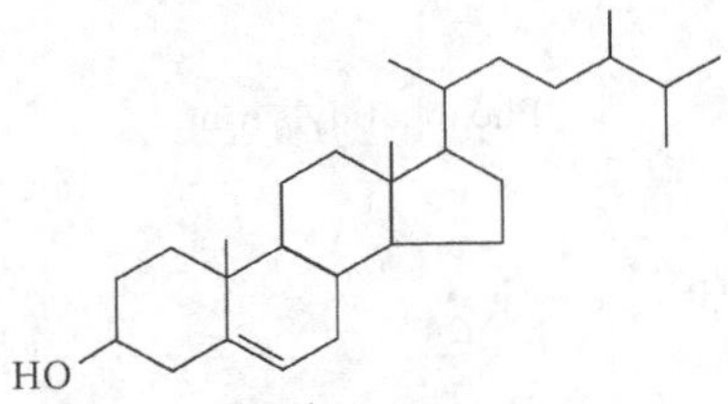

Campesterol

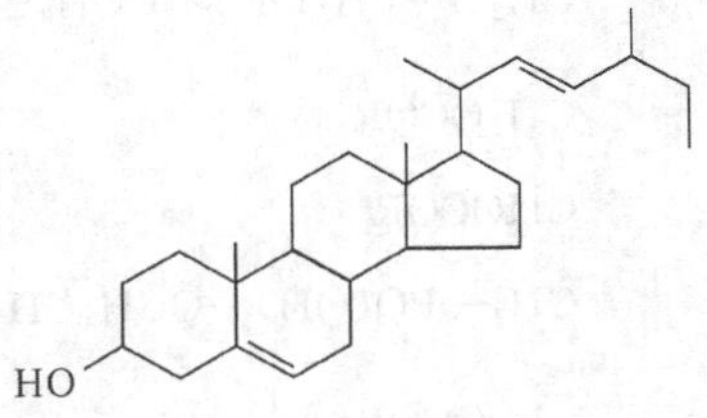

Brassicasterol

Fig. 5.6 Structures of the plant sterols

Table 5.13 Sterol content of fats and oils

Fat	Sterol(%)
Lard	0.12
Beef tallow	0.08
Milk fat	0.3
Herring	0.2-0.6
Cottonseed	1.4
Soybean	0.7
Corn	1.0
Rapeseed	0.4
Coconut	0.08
Cocoa butter	0.2

5.4.3 Phospholipids

All fats and oils and fat-containing foods contain a number of phospholipids. The lowest amounts of phospholipid are present in pure animal fats such as lard and beef tallow. In some crude vegetable oils, phospholipids may be present at levels of 2 to 3 percent. Fish, crustacea, and mollusks contain approximately 0.7 percent of phospholipids in the muscle tissue. Phospholipids are surface active, because they contain a lipophilic and hydrophilic portion. Since they can easily be hydrated, they can be removed from fats and oils during the refining process. The structures of the most important phospholipids are given in Fig. 5.7. After refining, neutralization, bleaching and deodorization of oils, the phospholipid content is reduced to virtually zero. The phospholipids removed from the soybean oil contain mainly lecithin and cephalin are used as emulsifiers in certain foods, such as chocolate.

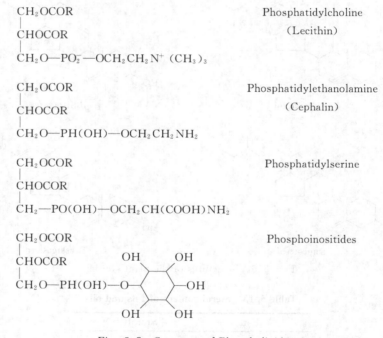

Fig. 5.7 Structure of Phospholipid

The distribution of fatty acids in phospholipids is not random, with saturated fatty acids preferentially occupying position 1 and unsaturated fatty acids position 2.

5.5 Fat Replacers and Mimetics

Fat replacers were developed in response to the consumer concern that quantities of lipid is included in modern diet and that has possible linkage with obesity. There are nearly 300 materials including modified starches, fiber, gums, emulsifiers, restructured protein, and cellulose proposed as fat replacers to play the roles of fats as ingredients in foods. Fat replacers are either fat substitutes or mimetics. Fat substitutes are lipid-like substances intended to replace fats on a one-to-one basis. Fat mimetics are protein or carbohydrate ingredients which function by imitating the physical, textural, mouthfeel, and organoleptic properties of real fats.

5.5.1 Fat Substitutes

The challenge for food technologists is to maintain optimal food texture and organoleptic properties whilst to reduce fat content. The fat substitutes are intended to replace the function of natural fats on a one-to-one weight basis. It is relatively easy to duplicate the emulsification of lipid using non-digestible surfactants in place of the surface active natural lipids such as phospholipids, monoglycerides, and diglycerides.

Sucrose esters with 6-8 moles C12-C22 fatty acids attached to sucrose have been approved in 1996 by America government for use in savoury foods. Less esterified sucrose fatty acid esters (2-3 moles of fatty acids) have been produced which are hydrophilic and also more digestible. Similar surfactants have been produced by attaching hydrophobic fatty acids to hydrophilic sugar alcohols and other polyols include tetrahalose, raffinose, and stachyose.

The medium chain triacylglycerides (MCTs) differ from natural fats by their relative absence of long-chain fatty acids. To produce MCTs coconut oil or palm kernel oil which are rich in medium chain fatty acids are first hydrolyzed and medium chain saturated fatty acids are isolated by fractionation. The medium chain fatty acid fraction is then reattached to glycerol with the aid of a catalyst. Compared to ordinary fats, MCTs have lower melting points, high solubility in water, and are resistant to oxidation. The MCTs are not transferred to the body's store of adipose tissue but are metabolized directly in the liver.

5.5.2 Fat Mimetics

The fat mimetics are nonlipid compounds that are able to simulate the physical functionalities of fats, namely creamy and smooth qualities, and usually are of carbohydrate or protein.

Carbohydrate fat mimetics include microparticulate cellulose and gelling agents. These materials provide the mouthfeel and flow properties of fat but are lacking the flavor characteristics associated with edible fats. Carbohydrate fat mimetics have been suggested as useful in dairy products, sauces, frozen desserts, and salad dressings. Carbohydrate fat mimetics have been classified as GRAS (generally regarded as safe) ingredients by the FDA (USA). Microparticulate cellulose also retains moisture and acts as a texturizer and a stabilizer. The second class of carbohydrate-based fat mimetics includes starch and modified starches. Their main functions are as bodying agents and texture modifiers intended to be used together with emulsifiers, proteins, gums, and other modified starches.

Protein fat mimetics are made from whey protein or milk and egg protein. The manufacturing process involves simultaneous heating and shearing to produce small particles of coagulated protein. This material also provides the mouthfeel of fat, could not be used for frying like other fat mimetics, but is stable for baking. This ingredient may be used in dairy products, salad dressing, margarine and mayonnaise-type products, baked foods, coffee creamer, soups, and sauces.

Glossary

2-dimethoxypropane	二甲氧基丙烷
abstracted	分离出来的
acetone	丙酮
acyl	酰基
adipose	动物脂肪
aldehyde	醛,乙醛
aliphatic	脂肪族的
ascorbyl palmitate	棕榈酸抗坏血酸酯
autoxidation	自动氧化
benzidine	联苯胺,对二氨基联苯
brassicasterol	菜籽甾醇
butylated hydroxy toluene	丁基羟基甲苯
butylated hydroxyanisole	叔丁基羟基茴香醚
campesterol	菜油甾醇
carbonyl	羰基
cardiovascular	心血管
carotenoid	类胡萝卜素
cephalin	脑磷脂
cerebroside	脑苷脂
cerebrovascular	脑血管
chloroform	氯仿
chlorophyll	叶绿素
constituent	成分
crustacea	甲壳动物

crystalline structure	晶体结构
culinary	烹调用的
customary	习惯的,通常的
deactivator	钝化剂,猝灭剂
deleterious	有毒的,有害的
depot fat	贮藏脂肪
derived	衍生的
diene	二烯
distinguish	区别
emulsifier	乳化剂
emulsion	乳液
erucic	芥子酸
ester	酯类
esterification	酯化作用
ethyl oleate	油酸乙酯
fluorescent	荧光
geometric	几何学的
glyceride	甘油酯
glycerol	甘油
grainy texture	沙粒质构
halogen	卤素
heterogeneous	多相的,非均匀的
hexane	己烷
hydrate	水合物
hydrocarbon	碳氢化合物,烃类
hydroperoxides	氢过氧化物
hydroquinine	氢化奎宁
hydroxyl	羟基
illuminated	照明
immiscible	不相溶的
induction period	诱导期
initiation	开始,起始
interconvertible	可相互转化
interesterification	酯交换
isomeric	同分异构的
isomerism	同分异构
ketone	酮
kink	蜷缩,扭结

lauric acid	月桂酸
lecithin	卵磷脂
linolenic acid	亚麻酸
lipoxidase	脂肪氧合酶
malabsorption	吸收不良
margarine	人造奶油
methanolysis	甲醇分解
methyl	甲基
methyl ester	甲酯
methylene	亚甲基
mimetics	模仿的
mollusks	软体动物
myoglobin	肌红蛋白
myristic	肉豆蔻酸
Na-methylate	甲醇钠
nonvolatile	非挥发性的
nuclei	原子核,核心
obesity	肥胖
objectionable	令人不愉快的,讨厌的
octanoic acid	辛酸
olefinic	烯族的
oleic-linoleic acid	油酸-亚油酸
organoleptic	感官的
organoleptic properties	感官特性
oxidative rancidity	氧化酸败
palatability	风味,适口性
palmitic	棕榈酸
peroxide	过氧化氢
peroxy-free radical	过氧自由基
petroleum ether	石油醚
phospholipid	磷脂
photosensitize	光猝灭
phytosterol	植物甾醇
polymorph	同分异构体
polymorphism	同质多晶,多型
prooxidant	促氧化
pro-oxidant	氧化剂,助氧化剂
propagation	延长,增殖

Propyl gallate	没食子酸丙酯
quinone	醌类
raffinose	棉子糖
rancid	腐臭的,讨厌的
refrigeration	制冷,冷藏
riboflavin	核黄素
ruminants	反刍动物
saponifier	皂化剂
saturation	饱和
savoury	可口的
sensitizer	激活剂
shortening	起酥油
sitosterol	谷甾醇
soya-lecithin	大豆卵磷脂
sphingolipid	鞘脂类
stachyose	水苏糖
stearodiolein	一硬脂酸二油酸甘油酯
stereospecific	立体定向的
steroid	类固醇
sterol	甾醇,固醇
stigmasterol	豆甾醇
subcell	亚晶胞
sulfolipid	硫酸脂
synergist	增效剂
tailor-made	特制的,量身定做的
tallow	动物油脂
termination	结束,终止
tert-butyl	叔丁基
thiosulfate	硫代硫酸盐
tocopherol	生育酚
triacyglycerol	甘油三酯
triglyceride	甘油三酯
triolein	三油酸甘油酯
tristearin	三硬脂酸甘油酯

Chapter 6   Vitamins

6.1   Introduction

Vitamins are minor but essential constituents of food. They are required for the normal growth, maintenance and functioning of the human body. Hence, their preservation during the storage and processing far-reaching importance. Data provided in Table 6.1 and illustrate vitamin losses in some preservation methods for fruits and vegetables. Vitamin losses can occur through chemical reactions which lead to inactive products, or by extraction or leaching, as in the case of water-soluble vitamins during blanching and cooking.

Table 6.1   Vitamin losses (%) through processing/canning of vegetables and fruits

Processed/ Canned product	Samples analyzed	Vitamin losses as % of freshly cooked and drained product				
		A	B_1	B_2	Niacin	C
Frozen products (cooked and drained)	10[a]	12[e] 10-50[f]	20 0-61	24 0-45	24 0-56	26 0-78
Sterilized products (drained)	7[b]	10 0-32	67 56-83	42 14-50	49 31-65	51 28-67
Frozen products (not thawed)	8[c]	37[e] 0-78[f]	29 0-66	17 0-67	16 0-33	18 0-50
Sterilized products (including the cooking water)	8[d]	39 0-68	47 22-67	57 33-83	42 25-60	56 11-86

[a] Asparagus, lima beans, green beans, broccoli, cauliflower, green peas, potatoes, spinach, brussels sprouts, and baby corn-cobs.

[b] As under with the exception of broccoli, cauliflower and brussels sprouts; the values for potato include the cooking water.

[c] Apples, apricots, bilberries, sour cherries, orange juice concentrate (calculated for diluted juice samples), peaches, raspberries and strawberries.

[d] As under except orange juice and not its concentrate was analyzed.

[e] Average values.

[f] Variation range.

The vitamin requirement of the body is usually adequately supplied by a balanced diet. A deficiency can result in hypovitaminosis and, if more severe, in avitaminosis. Both can occur not only as a consequence of insufficient supply of vitamins by food intake, but can be caused by disturbances in resorption, by stress and by disease. An assessment of the extent of vitamin supply can be made by determining the vitamin content in blood plasma, or by measuring a biological activity which is dependent on the presence of a vitamin, as many enzyme activities are. Vitamins are usually divided into two general classes: the fat-soluble vitamins, such as A, D, E and K_1, and the water-soluble vitamins, B_1, B_2, B_6, nicotinamide, pantothenic acid, biotin, folic acid, B_{12} and C. Data on the desirable human daily intake of some vitamins are presented by the age group in Table 6.2.

Table 6.2 Recommended daily intake of vitamins

Age group (years)	A (mg Retinol[a])	D (μg)[b]	E (mg)[c]	K (μg)[d]	C (mg)	B_1 (mg)	B_2 (mg)	Niacin[e] (mg)	B_6 (μg)	Folic acid[f] (mg)	Pantothenic acid (mg)	Biotin (μg)	B_{12} (μg)
<1	0.5-0.6	10	3-4	4-10	50-55	0.2-0.4	0.3-0.4	2-5	0.1-0.3	60-80	2-3	5-10	0.4-0.8
1-4	0.6	5	6	15	60	0.6	0.7	7	0.4	200	4	10-15	1.0
4-10	0.7-0.8	5	8-10	20-30	70-80	0.8-1.0	0.9-1.1	10-12	0.5-0.7	300	4-5	15-20	1.5-1.8
10-15	0.9-1.1	5	10-14	40-50	90-100	1.0-1.3	1.2-1.6	13-18	1.0-1.4	400	5-6	20-35	2.0-3.0
15-25	0.9-1.1	5	15	60-70	100	1.0-1.3	1.2-1.5	13-17	1.2-1.6	400	6	30-60	3.0
25-51	0.8-1.0	5	14	60-70	100	1.0-1.2	1.2-1.4	13-16	1.2-1.5	400	6	30-60	3.0
52-65	0.8-1.0	5	13	80	100	1.0-1.1	1.2-1.3	13-15	1.2-1.5	400	6	30-60	3.0
>65	0.8-1.0	10	12	80	100	1.0	1.2	13	1.2-1.4	400	6	30-60	3.0
Pregnant women	1.1	5	13	60	100	1.2	1.5	15	1.9	600	6	30-60	3.5
Lactating women	1.5	5	17	60	150	1.4	1.6	17	1.9	600	6	30-60	4.0

[a] 1mg retinol=1mg retinol equivalent=6mg all-trans-β-carotene=12mg provitamin A=1.15mg all trans-retinol acetate= 1.83mg all-trans-retinyl palmitate(IU=0.34μg retinol)

[b] Ergocalciferol (D2) or cholecalciferol (D3)(1 IU=0.025μg)

[c] Tocopherol equivalent (cf. 6.2.3.1)

[d] Phylloquinone (cf. 6.2.4)

[e] 1mg niacin equivalent=60 mg tryptophan

[f] 1μg folate equivalent=1 μg food folate=0.5 μfolic acid(PGA, cf. 6.3.7.1)

6.2 Fat-Soluble Vitamins

6.2.1 Vitamin A

6.2.1.1 Biological Roles

Retinol (I , in Formula 6.1) is important in protein metabolism of cells which develop from the ectoderm (such as skin or mucouscoated linings of the respiratory or digestive systems). Lack of retinol in some way negatively affects epithelial tissue (thickening of skin, hyperkeratosis) and also causes night blindness.

(Formula 6.1)

Furthermore, retinol, in the form of 11-cis-retinal (II), is the chromophore component of the visual cycle chromoproteins in three types of cone cells, blue, green and red (λ_{max} 435, 540 and 565 nm, respectively) and of rods of the retina.

The chromoproteins (rhodopsins) are formed in the dark from the corresponding proteins (opsins) and 11-cis-retinal, while in the light the chromoproteins dissociate into the more stable all-trans-retinal and protein. This conformational change triggers a nerve impulse in the adjacent nerve cell. The all-trans-retinal is then converted to all-trans-retinol and through an intermediary, 11-cis-retinol, is transformed back into 11-cis-retinal (see Fig. 6.1 for the visual cycle reactions).

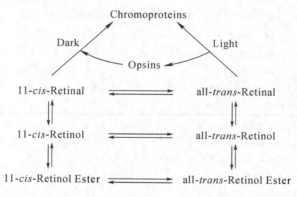

Fig. 6.1 Schematic representation of the visual cyle

6.2.1.2 Requirement, Occurrence

The daily requirement of vitamin A (Table 6.2) is provided to an extent of 75% by retinol intake (as fatty acid esters, primarily retinyl palmitate), while the rest of 25% is through β-carotene and other provitaminactive carotenoids. Due to the limited extent of carotenoid cleavage, at least 6 g of β-carotene are required to yield 1 g retinol. Vitamin A absorption and its storage in the liver occur essentially in the form of fatty acid esters. Its content in the liver is 250 μg/g fresh tissue, i.e., a total of about 240-540 mg is stored. The liver supplies the blood with free retinol, which then binds to proteins in blood. The plasma concentration of retinol averages 1.78 μmol/L in women and 2.04 μmol/L in men.

A hypervitaminosis is known, but the symptoms disappear if the intake of retinal is decreased. Vitamin A exist only in animal tissues; expercially in fish liver oil, livers of mammals, milk fat and egg yolk. Plants are devoid of vitamin A but do contain carotenoids which yield vitamin A by breaking of the centrally located double bond (provitamins A). Carotenoids are present in almost all vegetables but primarily in green, yellow and leafy vegetables (carrots, spinach, cress, kale, bell peppers, paprika peppers, and tomatoes) and in fruit, outstanding sources being rose hips, pumpkin, apricots, oranges and palm oil, which is often used for yellow colorings. Animal carotenoids are always of plant origin, derived from the feed. Table 6.4 gives the vitamin A content of some common foods. These values can vary greatly with cultivar, stage of ripeness, etc. An accurate estimate of the vitamin A content in food must include a detailed analysis of its carotenoids.

6.2.1.3 Stability, Degradation

Food processing and storage can lead to 5%-40% destruction of vitamin A and carotenoids. In the absence of oxygen and at high temperature, as experienced in cooking or food sterilization, the main reactions are isomerization and fragmentation. In the presence of oxygen, oxidative degradation leads to a series of products, some of which are volatile (cf. 3.8.4.4). This oxidation often parallels lipid oxidation (cooxidation process). The rate of oxidation is influenced by oxygen partial pressure, water activity, temperature, etc. Dehydrated foods are particularly sensitive to the oxidative degradation.

6.2.2 Vitamin D

6.2.2.1 Biological Role

Cholecalciferol (vitamin D_3, Ⅰ) is formed from cholesterol in the skin through photolysis of 7-dehydrocholesterol (provitamin D_3) by ultra-violet light ("sunshine vitamin"; cf. 3.8.2.2.2). Similarly, vitamin D_2 (ergocalciferol, Ⅱ; cf. Formula 6.2) is formed from the ergosterol.

Vitamin D_2 and D_3 are hydroxylated first in the liver to the prohormone 25-hydroxychole-calciferol (calcidiol) and subsequently in the kidney to the vitamin D hormone 1α, 25-dihy-droxycholecalciferol (calcitriol). Calcitriol acts as an inductor of proteins in various organs. It promotes calcium resorption in the intestine and helps to reach an optimal calcium concentration in the kidney and bones, inducing the synthesis of proteins involved in the structure of the bone matrix and calcification.

Vitamin D deficiency can result in an increased excretion of calcium and phosphate and, consequently, impairs bone formation through the inadequate calcification of cartilage and bones (childhood rickets). Vitamin D deficiency in adults leads to osteomalacia, a softening and weakening of bones. Hypercalcemia is a result of excessive intake of vitamin D (>50 μg/day), causing calcium carbonate and calcium phosphate deposition disorders involving various organs.

(Formula 6.2)

6.2.2.2 Requirement, Occurrence

The daily requirement is shown in Table 6.2. Indicators of deficiency are the concentration of the metabolite 25-hydroxycholecalciferol in plasma and the activity of alkaline serum phosphatase, which increases during the vitamin deficiency. Most natural foods have a low content of vitamin D_3. Fish liver oil is an exceptional source of vitamin D_2. The D-provitamins, ergosterol and 7-dehydrocholesterol, are widely distributed in the animals and plant kingdoms. Yeast, some mushrooms, cabbage, spinach and wheat germ oil are particularly rich in provitamin D_2. Vitamin D_3 and its provitamin are present in egg yolk, butter, cow's milk, beef and pork liver, mollusks, animal fat and pork skin. However, the most important vitamin D source is fish oil, primarily liver oil. The vitamin D requirement of humans is best supplied by 7-dehydrocholesterol. Table 6.4 gives data on vitamin D occurrence in some the foods. However, these values can vary widely, as shown by variations in dairy cattle milk (summer or winter), caused by feed or frequency of pasture grazing and exposure to the ultraviolet rays of sunlight.

6.2.2.3 Stability, Degradation

Vitamin D is sensitive to oxygen and light. It is relatively stabile in foods.

6.2.3 Vitamin E

6.2.3.1 Biological Role

The various tocopherols differ in the number and position of the methyl groups on the ring. α-Tocopherol (Formula 6.3; the configuration at the three asymmetric centers, 2, 4 and 8, is R) has the highest biological activity. Its activity is based mainly on its antioxidative properties, which slows down of prevents lipid oxidation (cf. 5.3). Thus, it not only contributes to the stabilization of membrane structures, but also stabilizes other active agents (e.g., vitamin A, ubiquinone, hormones, and enzymes) against oxidation. Vitamin E is involved in the conversion of arachidonic acid to prostaglandins and slows down the aggregation of blood platelets. Vitamin E deficiency is associated with chronic disorders (sterility in domestic and experimental animals, anemia in monkeys, and muscular dystrophy in chickens). Its mechanism is not fully elucidated.

(Formula 6.3)

6.2.3.2 Requirement, Occurrence

The daily requirement is given in Table 6.2. It increases when the diet contains a high content of unsaturated fatty acids. A normal supply results in a tocopherol concentration of 12-46 μmol/L in the blood plasma.

Table 6.4 provides data on the tocopherol content in some foods. The main sources are vegetable oils, especially germ oils of cereals.

6.2.3.3 Stability, Degradation

Losses happen in vegetable oil processing into margarine and shortening. Losses are also encountered in the intensive lipid autoxidation, particularly in dehydrated or deep fried foods (Table 6.3).

Table 6.3 Tocopherol stability during deep frying

	Tocopherol total(mg/100 g)	Loss(%)
Oil before deep frying	82	
after deep frying	73	11
Oil extracted from potato chips		
immediately after production	75	
after 2 weeks storage at room temperature	39	48
after 1 month storage at room temperature	22	71
after 2 month storage at room temperature	17	77
after 2 months kept at -12°C	28	63
after 1 months kept at -12°C	24	68
Oil extracted from French fries		
immediately after production	78	
after 1 months kept at -12°C	25	68
after 2 months kept at -12°C	20	74

6.2.4 Vitamin K

6.2.4.1 Biological Roles

The K-group vitamins are naphthoquinone derivatives which differ in their side chains. The structure of vitamin K_1 is shown in (Formula 6.4) The configuration at carbon atoms 7' and 11' is R and corresponds to that of natural phytol. Racemic vitamin K_1 synthesized from the optically inactive isophytol has the same biological activity as the natural product. Vitamin K is involved in the post-translational synthesis of γ-carboxyglutamic acid (Gla) in vitamin K-dependent proteins. It is reduced to the hydroquinone form (Formula 6.4) which acts as a cofactor in the carboxylation of glutamic acid. In this process, it is converted to the epoxide from which vitamin K is regenerated. Blood clotting factors (prothrombin, proconvertin, Christmas and Stuart factor) as well as proteins which perform other functions belong to the group of vitamin K-dependent proteins which bind Ca^{2+} ions at Gla. Deficiency of this vitamin causes the reduction of prothrombin activity, hypothrombinemia and hemorrhage.

(Formula 6.4)

Red.: Reductase; Carb.: Carboxylase

6.2.4.2 Requirement, Occurrence

The activity is given in vitamin equivalents (VE): 1 VE = 1 μg phylloquinone. The daily requirement of vitamin K_1 is shown in Table 6.2. It is covered by food (cf. Table 6.4). The bacteria present in the large intestine form relatively high amounts of K_2. However, it is uncertain whether they appreciably contribute to covering the requirement.

Vitamin K_1 occurs primarily in green leafy vegetables (spinach, cabbage, and cauliflower), and liver (veal or pork) is also an excellent source (Table 6.4).

Chapter 6 Vitamins

Table 6.4 Vitamin content of some food products[a]

Food product	Carotene mg	A mg	D μg	E mg	K mg	B₁ mg	B₂ mg	NSM[c] mg	PAN[d] mg	B₆ mg	BIO[e]	FOL[f]	B₁₂	C mg
Milk and milk products														
Bovine milk	0.018	0.028	0.088	0.07	0.0003	0.04	0.18	0.09	0.35	0.04	3.5	8.0	0.4	1.7
Human milk	0.003	0.054	0.07	0.28	0.0005	0.02	0.04	0.17	0.21	0.01	0.6	8.0	0.05	6.5
Butter	0.38	0.59	1.2	2.2	0.007	0.005	0.22	0.03	0.05	0.005				0.2
Cheese														
Emmental	0.12	0.27	1.1	0.53	0.003	0.05	0.34	0.18	0.40	0.11	3.0	9.0	3.0	0.5
Camembert(60% fat)	0.29	0.50		0.77		0.04	0.37	0.95	0.7	0.2	2.8	38	2.4	
Camembert(30% fat)	0.1	0.2	0.17	0.30		0.05	0.67	1.2	0.9	0.3	5.0	66	3.1	
Eggs														
Chicken egg yolk	0.29	0.88	5.6	5.7		0.29	0.40	0.07	3.7	0.3	53	208	2.0	0.3
Chicken egg white						0.02	0.32	0.09	0.14	0.012	7	9.2	0.1	0.3
Meat and meat products		28.0												
Beef, muscle		0.02		0.48	0.013	0.08	0.26	7.5	0.31	0.24	3.0	3.0	5.0	
Pork, muscle		0.006		0.41	0.018	0.90	0.23	5.0	0.70	0.4		0.8		
Calf liver		28.0	0.33	0.24	0.09	0.28	2.61	15.0	7.9	0.17	80	240	60	35
Pork liver		36		0.60	0.06	0.31	3.2	15.7	6.8	0.6	30	220	40	23
Chicken liver		33	1.3	0.50	0.08	0.32	2.49	11.6	7.2	0.8		380	20	28
Pork kidney		0.06		0.45		0.34	1.8		3.1	0.6			20	16
Blood sausage	0.02					0.07	0.13						50	
Fish and fish products														
Herring		0.04	27	1.5		0.04	0.22		0.9	0.5	4.5	5	8.5	0.5
Eel		0.98	20	8		0.18	0.32		0.3	0		13	1	1.8
Cod-live oil				3.26										
Cereals and cereals product														
Wheat, whole kernel	0.02			1.4		0.48	0.09		1.2	0.27	6	58		
Wheat flour, type 550				0.34		0.11	0.03		0.4	0.10	1.1	16		
Wheat flour, type1050						0.43	0.07		0.63	0.24	1.1	30		
Wheat jerm				27.6	0.13	2.01	0.72		1.0	0.5	17	520		
Rye whole kernel	0.06			2.0		0.35	0.17		1.5	0.23	5.0	35		
Rye flour type997						0.19	0.11					33		
Corn whole kernel	13			2.0	0.04	0.36	0.20		0.7	0.40	6	31		
Corn(breakfast cereal, corn flakes)	0.17			0.18		0.6		1.4	0.2	0.07		6		
Oat flakes				1.5	0.063	0.59	0.15	1.0	1.1	0.16	20	67		
Rice, unpolished				0.74		0.41	0.09	5.2	1.7	0.28	12	16		
Rice, polished				0.18		0.06	0.03	1.3	0.6	0.15	3.0	11		
Vegetables														
Watercress	0.49					0.09	0.17	0.7						96
Mushrooms(cultivated)	0.01		1.94	0.12	0.02	0.10	0.44	5.2	2.1	0.07	16	25		4.9
Chicory	3.4					0.06	0.03	0.24		0.05	4.8	50		8.7

(To be continued)

FOOD CHEMISTRY

(Table 6.4)

Food product	Carotene mg	A mg	D µg	E mg	K mg	B_1 mg	B_2 mg	NSM[c] mg	PAN[d] mg	B_6 mg	BIO[e]	FOL[f]	B_{12}	C mg
Endive	1.7					0.05	0.12	0.4				109		9.4
Lamb's lettuce	3.9			0.6		0.07	0.08	0.4		0.25		145		35
Kale	5.2			1.7	0.82	0.1	0.25	2.1		0.3	0.5	187		105
Potatoes	0.005			0.05	0.002	0.110	0.05	1.2	0.4	0.31	0.4	15		17
Kohlrabi	0.2					0.05	0.05	1.8	0.1	0.1	2.7	70		63
Head lettuce	1.1			0.6	0.2	0.06	0.08	0.3	0.1	0.06	1.9	41		13
Lentils, dried	0.1					0.48	0.26	2.5	1.4	0.6		168		7.0
Carrots	12			0.47	0.015	0.07	0.05	0.6	0.3	0.27	5	17		7.1
Brussels sprouts	0.4			0.6	0.24	0.13	0.14	0.7		0.3	0.4	101		114
Spinach	4.8			2.5	0.4	0.09	0.20	0.6	0.3	0.22	6.9	145		52
Edible mushroom (*Boletus edulis*)			3.1	0.63		0.03	0.37	4.9	2.7					2.5
Tomatoes	0.59			0.81	0.006	0.06	0.04	0.5	0.3	0.1	4	33		19
White cabbage	0.07			1.7	0.07	0.05	0.04	0.3	0.3	0.19	3.1	31		48
Fruits														
Apple	0.05			0.49	0.004	0.035	0.032	0.3	0.1	0.1	0.0045	5		12
Orange	0.1			0.32		0.08	0.04	0.3	0.2	0.1	2.3	22		50
Apricot	1.8			0.5		0.04	0.05	0.8	0.3	0.1		4		9.4
Strawberry	0.02			0.12	0.02	0.03	0.05	0.5	0.3	0.06	4	43		64
Grapefruit	0.01			0.30		0.05	0.02	0.24	0.25	0.03	0.4	10		44
Rose hips	4.8			4.2		0.09	0.06	0.48		0.05				1250
Red currant	0.03			0.71		0.04	0.03	0.23	0.06	0.05	2.6	11		36
Black currant	0.08			1.9		0.05	0.04	0.28	0.4	0.08	2.4	8.8		177
Sour cherries	0.24			0.13		0.05	0.06	0.4				29		12
Plums	0.41			0.86		0.07	0.04	0.4	0.2	0.05	0.1	2		5.4
Sea buckthorn	1.5			3.2		0.03	0.21	0.3	0.2	0.11	3.3	10		450
Yeast														
Baker's yeast, pressed						1.43	2.31	17.4	3.5	0.68	33	293		
Brewer's yeast, dried						12.0	3.8	44.8	7.2	4.4	20			

[a] Values are given in mg or µg per 100g of edible portion. [b] Total carotenoids with vitamin A activity. [c] Nicotinamide. [d] Pantothenic acid. [e] Folic acid.

6.2.4.3 Stability, Degradation

Little is known about the reactions of vitamin K_1 in foods. The vitamin K compounds are destroyed by light and alkali. They are relatively stable to atmospheric oxygen and exposure to heat. In the hydrogenation of oils, the double bond in residue R (cf. Formula 6.4) is attacked. Although hydrogenated vitamin K (2,3-dihydrophylloquinone) is absorbed, it is apparently no longer as active as the natural form.

6.3 Water-Soluble Vitamins

6.3.1 Ascorbic Acid

6.3.1.1 Biological Role

Ascorbic acid (L-3-keto-threo-hexuronic acid-γ-lactone) is involved in hydroxylation reactions, e.g., biosynthesis of catecholamines, hydroxyproline and corticosteroids (11-β-hydroxylation of deoxycorticosterone and 17-β-hydroxylation of corticosterone). Vitamin C is fully absorbed and distributed throughout the body, with the highest concentration in adrenal and pituitary glands.

About 3% of the body's vitamin C pool, which is 20-50 mg/kg body weight, is excreted in the urine as ascorbic acid, dehydroascorbic acid (a combined total of 25%) and their metabolites, 2,3-diketo-L-gulonic acid (20%) and oxalic acid (55%). An increase in excreted oxalic acid occurs only with a very high intake of ascorbic acid. Scurvy is caused by a dietary deficiency of ascorbic acid.

6.3.1.2 Requirement, Occurrence

The daily requirement is shown in Table 6.2. An indicator of insufficient vitamin supply in the diet is a low level in blood plasma (0.65 mg/100 ml). Vitamin C is present in all animals and plant cells, mostly in free form, and it is probably bound to protein as well. Vitamin C is particularly abundant in rose hips, black and red currants, strawberries, parsley, oranges, lemons (in peels more than in pulp), grapefruit, a variety of cabbages and potatoes. Vitamin C loss during storage of vegetables from winter to late spring can be as high as 70%.

Table 6.4 provides data on vitamin C occurrence in a variety of foods.

Ascorbic acid is chemically synthesized. However, the synthesis by means of genetically modified microorganisms (GMO vitamin C) is more cost effective. Therefore, the largest proportion is synthesized by these means.

6.3.1.3 Stability, Degradation

Ascorbic acid (I) has an acidic hydroxyl group ($pK_1 = 4.04$, $pK_2 = 11.4$ at 25 °C). Its UV absorption depends on the pH value (Table 6.5). Ascorbic acid is readily and reversibly oxidized to dehydroascorbic acid (II), which is present in aqueous media as a hydrated hemiketal (IV). The biological activity of II is possibly weaker than that of I because the plasma and tissue concentrations of II are considerably lower after the administration of equal amounts of I and II. The activity is completely lost when the dehydroascorbic acid lactone ring is irreversibly opened, converting II to 2,3-

diketogulonic acid (Ⅲ), cf. Formula 6.5:

(Formula 6.5)

The oxidation of ascorbic acid is dehydroascorbic acid and its further degradation products depends on a number of parameters. Oxygen partial pressure, pH, temperature and the presence of heavy metal ions are of great importance. Metal-catalyzed destruction proceeds at a higher rate than noncatalyzed spontaneous autoxidation. Traces of heavy metal ions, particularly Cu^{2+} and Fe^{3+}, result in high losses.

Table 6.5 Effect of pH on ultraviolet absorption maxima of ascorbic acid

pH	λ_{max}
2	244
6-10	266
>10	294

The principle of metal catalysis is schematically presented in reaction (Formula 6.6).

$$\text{(Formula 6.6)}$$

The rate of anaerobic vitamin C degradation, which is substantially lower than that of non-catalyzed oxidation, is maximal at pH 4 and minimal at pH 2. It probably proceeds through the ketoform of ascorbate, then via a ketoanion to diketogulonic acid:

$$R-CO-CO-COOH \quad \text{(Formula 6.7)}$$

Diketogulonic acid degradation products, xylosone and 4-deoxypentosone (Formula 6.7) are then converted into ethylglyoxal, various reduc-tones (Formula 6.8), furfural and furancarboxylic acid.

$$\text{(Formula 6.8)}$$

FOOD CHEMISTRY

(Formula 6.9)

In the presence of amino acids, ascorbic acid, dehydroascorbic acid and their degradation products might be changed further by entering into Maillard-type browning reactions (Formula 6.9). An example is the reaction of dehydroascorbic acid with amino compounds to give pigments, which can cause unwanted browning in citrus juices and dried fruits. The intermediates that have been identified are scorbamic acid (I in Formula 6.9), which is produced by Strecker degradation with an amino acid, and a red pigment (II). A wealth of data is available on ascorbic acid losses during preservation, storage and processing of food. Tables 6.1 and Fig. 6.2 and 6.3 present several examples. Ascorbic acid degradation is often used as a general indicator of changes occurring in food.

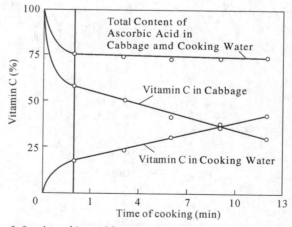

Fig. 6.2 Ascorbic acid losses as a result of cooking of cabbage

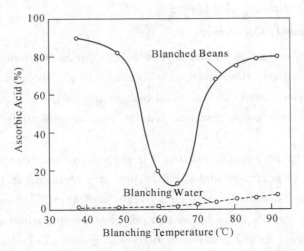

Fig. 6.3 Ascorbic acid losses in green beans versus blanching temperature

6.3.2 Thiamine (Vitamin B_1)

6.3.2.1 Biological Role

(Formula 6.10)

Thiamine, in the form of its pyrophosphate (cf. Formula 6.10), such as pyruvate dehydrogenase, transketolase, phosphoketolase and α-ketoglutarate dehydrogenase, in reactions involving the transfer of an activated aldehyde unit (Formula 6.11 D: donor; A: acceptor).

Formula 6.11

Vitamin B_1 deficiency is shown by a decrease in activity of the enzymes mentioned above. The disease beri-beri, which has neurological and cardiac symptoms, results from

FOOD CHEMISTRY

the severe dietary deficiency of thiamine.

6.3.2.2 Requirement, Occurrence

The daily requirement is shown in Table 6.2. Since thiamine is a key substance in carbohydrate metabolism, the requirement increases in a carbohydrate-enriched diet. The assay of transketolase activity in red blood cells or the extent of transketolase reactivation on addition of thiamine pyrophosphate can be used as indicators for sufficient vitamin intake in the diet.

Vitamin B_1 is found in many plants. It is present in the pericarp and germ of cereals, in yeast, vegetables (potatoes) and shelled fruit. It is abundant in pork, beef, fish, eggs and in animal organs such as liver, kidney, brain and heart. Human milk and cow's milk contain vitamin B_1. Whole grain bread and potatoes are important dietary sources. Since vitamin B_1 is localized in the outer part of cereal grain hulls, flour milling with a low extraction grade or rice polishing removes most of the vitamin in the bran (cf. 15.3.1.3 and 15.3.2.2.1). Table 6.4 lists data on the occurrence of thiamine.

6.3.2.3 Stability, Degradation

The stability of thiamine in aqueous solution is relatively low. It is influenced by pH (Fig. 6.4), temperature (Table 6.6), ionic strength and metal ions. The enzyme-bound form is less stable than free thiamine (Fig. 6.4). Strong nucleophilic reagents, such as HSO_3^- or OH^-, cause rapid decomposition by forming 5-(2-hydroxyethyl)-4-methylthiazole and 2-methyl-4-amino-5(methyl-sulfonic acid)-pyrimidine, or 2-methyl-4-amino-5-hydroxymethylpyrimidine (Formula 6.12).

(Formula 6.12)

Thermal degradation of thiamine, which also initially yields the thiazole and pyrimidine derivatives mentioned above, is involved in the formation of meat-like aroma in cooked food (cf. 5.3.1.4).

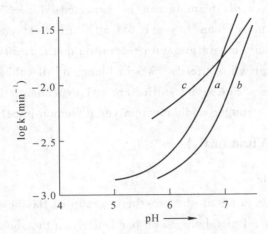

Fig. 6.4 Inactivation rate of thiamine as affected by pH

Table 6.6 **Thiamine losses in food during storage (12 months)**

	Thiamine loss, %	
	1.5 ℃	38 ℃
Apricots	28	65
Orange juice	0	22
Peas	0	32
Green beans	24	92
Tomato juice	0	40

a: Thiamine in phosphate buffer, b: thiamine in wheat or oat flour, c: thiamine pyrophosphate in flour

Thiamine is inactivated by nitrites, probably through the reaction with the amino group attached to the pyrimidine ring. Strong oxidants, such as H_2O_2 or potassium ferricyanide, yield the fluorescent thiochrome. This reaction is often used in the chemical determination of the thiamine the content in food (see reaction, Formula 6.13).

(Formula 6.13)

FOOD CHEMISTRY

The following losses of thiamine can be expected: 15%-25% in canned fruit or vegetables stored for more than a year; 0%-60% in meat cooked under household conditions, depending on temperature and preparation method; 20% in salt brine pickling of meat and in baking of white bread; 15% in blanching of cabbage without sulfite and 40% with sulfite. Losses caused by sulfite are pH dependent. Practically no thiamine degradation occurs in a stronger acidic medium (e. g. lemon juice).

6.3.3 Riboflavin (Vitamin B_2)

6.3.3.1 Biological Role

Riboflavin (Formula 6.14) is the prosthetic group of flavine enzymes, which are of great importance in general metabolism and particularly in metabolism of protein.

(Formula 6.14)

Riboflavin deficiency will lead to accumulation of amino acids. A specific deficiency symptom is the decrease of glutathione reductase activity in red blood cells.

6.3.3.2 Requirement, Occurrence

The daily requirement is given in Table 6.2. Deficiency symptoms are rarely observed with a normal diet and, since the riboflavin pool in the body is very stable, even in a deficient diet it is not depleted by more than 30%-50%. The riboflavin content of urine is an indicator of riboflavin supply levels. Values above $80\mu g$ riboflavin/g creatinine are normal; $27-79\mu g/g$ are low; and less than $27\mu g/g$ strongly suggests a vitamin-deficient diet. Glutathione reductase activity assay can provide similar information.

The most important sources of riboflavin are milk and milk products, eggs, various vegetables, yeast, meat products, particularly variety meats such as heart, liver and kidney, and fish liver and roe. Table 6.4 provides data about the occurrence of riboflavin in some common foods.

6.3.3.3 Stability, Degradation

Riboflavin is relatively stable in normal food handling processes. Losses range from 10%-15%. Exposure to light, especially in the visible spectrum from 420 to 560 nm, photolytically cleaves ribitol from the vitamin, converting it to lumiflavin:

6.3.4 Nicotinamide (Niacin)

6.3.4.1 Biological Role

Nicotinic acid amide (I), in the form of nicotinamide adenine dinucleotide (NAD$^+$, cf. 2.3.1.1), or its phosphorylated form (NADP$^+$), is a coenzyme of dehydrogenases. Its excretion in urine is essentially in the form of N^1-methylnicotinamide (trigonellinenamide, II), N^1-methyl-6-pyridone-3-carboxamide (III) and N^1-methyl-4-pyridone-3-carboxamide (IV):

(Formula 6.15)

Vitamin deficiency is observed initially by a drop in concentration of NAD$^+$ and NADP$^+$ in liver and muscle, while levels remain normal in blood, heart and kidney. The classical deficiency disease is pellagra, which affects the skin, digestion and the nervous system (dermatitis, diarrhea and dementia). However, the initial deficiency symptoms are largely non-specific.

6.3.4.2 Requirement, Occurrence

The daily requirement (cf. Table 6.2) is covered to an extent of 60%-70% by tryptophan intakes. Hence, milk and eggs, though they contain little niacin, are good foods for the prevention of pellagra because they contain tryptophan. It substitutes for niacin in the body, with 60 mg L-tryptophan equalling 1 mg nicotinamide. Indicators of sufficient supply of niacin in the diet are the levels of metabolites II (cf. Formula 6.15) in urine or III and IV in the blood plasma.

The vitamin occurs in food as nicotinic acid, either as its amide or as a coenzyme. Animal organs, such as liver, lean meat, cereals, yeast and mushrooms are abundant sources of niacin. Table 6.4 provides data on its occurrence in food.

6.3.4.3 Stability, Degradation

Nicotinic acid is quite stable. Moderate losses of up to 15% are observed (cf. Tables 6.1) in blanching of vegetables. The loss is 25%-30% in the first days of ripening of meat.

6.3.5 Pyridoxine (Pyridoxal, Vitamin B$_6$)

6.3.5.1 Biological Role

Vitamin B$_6$ activity is exhibited by pyridoxine (Formula 6.16) or pyridoxol (R=

CH_2OH), pyridoxal (R=CHO) and pyridoxamine (R=CH_2NH_2). The metabolically active form, pyridoxal phosphate, functions as a coenzyme of amino acid decarboxylases, amino acid racemases, amino acid dehydrases, amino transferases, serine palmitoyl transferase, lysyl oxidase, δ-aminolevulinic acid synthase, and of enzymes of tryptophan metabolism. Furthermore, it stabilizes the conformation of phosphorylases.

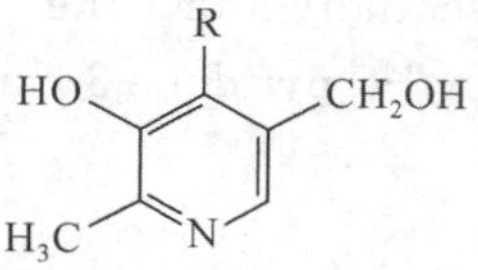

(Formula 6.16)

The intake of the vitamin occurs usually in the forms of pyridoxal or pyridoxamine.

Pyridoxine deficiency in the diet causes disorders in protein metabolism, e. g., in hemoglobin synthesis. Hydroxykynurenine and xanthurenic acid will accumulate because the conversion of tryptophan to nicotinic acid, a step regulated by the kynureninase enzyme, is interrupted.

6.3.5.2 Requirement, Occurrence

The daily requirement is given in Table 6.2. An indicator of sufficient supply is the activity of glutamate oxalacetate transaminase, an enzyme present in red blood cells. This activity is decreased in vitamin deficiency. The occurrence of pyridoxine in food is outlined in Table 6.4.

6.3.5.3 Stability, Degradation

The most stable form of the vitamin is pyridoxal, and this form is used for vitamin fortification of food. Vitamin B_6 loss is 45% in cooking of meat and 20%-30% in cooking of vegetables. During milk sterilization, a reaction with cysteine transforms the vitamin into an inactive thiazolidine derivative (Formula 6.17). This reaction may account for vitamin losses also in other heat-treated foods.

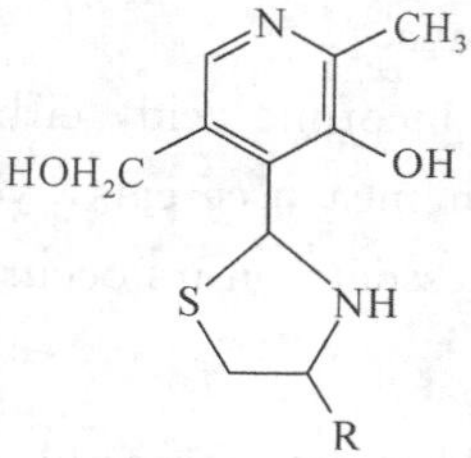

(Formula 6.17)

6.3.6 Folic Acid

6.3.6.1 Biological Role

The tetrahydrofolate derivative (Formula 6.18, II) of folic acid (I, pteroylmonoglutamic acid, PGA) is the cofactor of enzymes which transfer single carbon units in various oxidative states, e. g., formyl or hydroxymethyl residues. In transfer reactions the single carbon unit is attached to the N^5- or N^{10}-atom of tetrahydrofolic acid.

$$\text{(Formula 6.18)}$$

Folic acid deficiency caused by insufficient supply in the diet or by malfunction of absorptive processes is detected by a decrease in folic acid concentration in red blood cells and plasma, and by a change in blood cell patterns. There are clear indications that a congenital defect (neural tube defect) and some diseases are based on a deficiency of folate.

6.3.6.2 Requirement, Occurrence

The requirement shown in Table 6.2 is not often reached. In some countries, cereal products are supplemented with folic acid in order to avoid deficiency, e.g., with 1.4 mg/kg in the USA. Correspondingly, positive effects on consumer health have been observed.

In cooperation with vitamin B_{12}, folic acid methylates homocysteine to methionine. Therefore, homocysteine is a suitable marker for the supply of folate. In the case of a deficiency, the serum concentration of this marker is clearly raised compared with the normal value of 8-10 μmol/mL, resulting in negative effects on health because higher concentrations of homocysteine are toxic.

In food folic acid is mainly bound to oligo-γ-L-glutamates made up of 2-6 glutamic acid residues. Unlike free folic acid, the absorption of this conjugated form is limited and occurs only after the glutamic acid residues are cleaved by folic acid conjugase, a γ-glutamyl-hydrolase, to give the monoglutamate compound. Since certain constituents can reduce the absorption of folates, the average bioavailability is estimated at 50%. The folic acid content of foods varies. Data on folic acid occurrence in food are compiled in Table 6.4.

6.3.6.3 Stability, Degradation

Folic acid is quite stable. There is no destruction during blanching of vegetables, while cooking of meat gives only small losses. Losses in milk are apparently due to an oxidative process and parallel those of ascorbic acid. Ascorbate added to food preserves folic acid.

6.3.7 Biotin

6.3.7.1 Biological Role

Biotin (Formula 6.19) is the prosthetic group of carboxylating enzymes, such as acetyl-CoA-carboxylase, pyruvate carboxylase and propionyl-CoA-carboxylase, and therefore plays an important role in fatty acid biosynthesis and in gluconeogenesis. The carboxyl group of biotin forms an amide bond with the ε-amino group of a lysine residue of the particular enzyme protein. Only the (3aS, 4S, 6aR) compound, D-(+)-biotin, is biologically active:

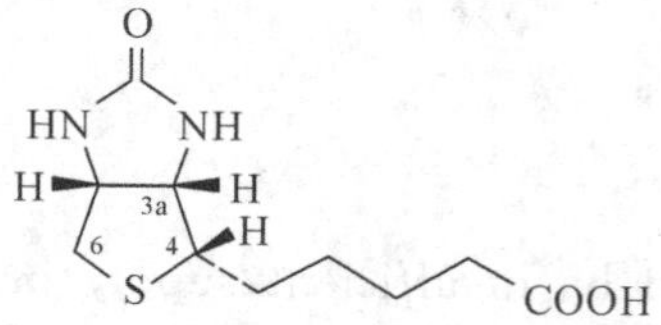

(Formula 6.19)

Biotin deficiency rarely occurs. Consumption of large amounts of raw egg white might inactivate biotin by its specific binding to avidin.

6.3.7.2 Requirement, Occurrence

The daily requirement is shown in Table 6.2. An indicator of sufficient biotin supply is the excretion level in the urine, which is normally 30-50 μg/day. A deficiency is indicated by a drop to 5 μg/day. Biotin is not free in food, but is bound to proteins. Table 6.4 provides data on its occurrence in food.

6.3.7.3 Stability, Degradation

Biotin is quite stable. Losses during processing and storage of food are 10%-15%.

6.3.8 Pantothenic Acid

6.3.8.1 Biological Role

Pantothenic acid (Formula 6.20) is the building unit of coenzyme A (CoA), the main carrier of acetyl and other acyl groups in cell metabolism. Acyl groups are linked to CoA by a thioester bond. Pantothenic acid occurs in free form in blood plasma, while in organs it is present as CoA. The highest concentrations are in liver, adrenal glands, heart and kidney.

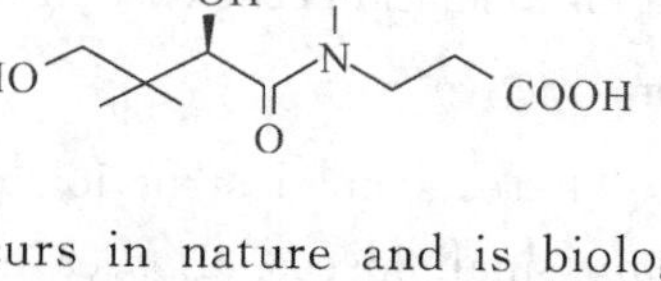

(Formula 6.20)

Only the R enantiomer occurs in nature and is biologically active. A normal diet provides an adequate supply.

6.3.8.2 Requirement, Occurrence

The daily requirement is 6-8 mg. The concentration in blood is 10-40 μg/100 mL and 2-7 mg/day are excreted in urine. Pantothenic acid in food is determined with microbiological or ELISA techniques. A gas chromatographic method using a ^{13}C-isotopomer of pantothenic acid as the internal standard is very accurate and much more sensitive. Table 6.4 lists data on pantothenic acid occurrence in food.

6.3.8.3 Stability, Degradation

Pantothenic acid is quite stable. Losses of 10% are experienced in processing of milk. Losses of 10%-30%, mostly due to leaching, occur during cooking of vegetables.

6.3.9 Cyanocobalamin (Vitamin B_{12})

6.3.9.1 Biological Role

Cyanocobalamin (Formula 6.21) was isolated in 1948 from Lactobacillus lactis. Due to its stability and availability, it is the form in which the vitamin is used most often. In fact, cyanocobalamin is formed as an artifact in the processing of biological materials. Cobalamins occur naturally as adenosylcobalamin and methylcobalamin, which instead of the cyano group contain a 5-deoxyadenosyl residue and a methyl group respectively.

Adenosylcobalamin (coenzyme B_{12}) participates in rearrangement reactions in which a hydrogen atom and an alkyl residue, an acyl group or an electronegative group formally exchange places on two neighboring carbon atoms. Reactions of this type play a role in the metabolism of a series of bacteria. In mammals and bacteria a rearrangement reaction that depends on vitamin B_{12} is the conversion of methylmalonyl CoA to succinyl CoA (cf. 10.2.8.3). Vitamin B_{12} deficiency results in the excretion of methylmalonic acid in the urine.

Another reaction that depends on adenosylcobalamin is the reduction of ribonucleoside triphosphates to the corresponding 2-deoxy compounds, the building blocks of-bonucleic acids.

Methylcobalamin is formed, e.g., in the methylation of homocysteine to methionine with N^5-methyltetrahydrofolic acid as the intermediate stage. The enzyme involved is a cobalamin-dependent methyl transferase.

The absorption of cyanocobalamin is achieved with the aid of a glycoprotein, the "intrinsic factor" formed by the stomach mucosa. Deficiency of vitamin B_{12} is usually caused by the impaired absorption due to the inadequate formation of "intrinsic factor" and results in pernicious anemia.

6.3.9.1 Requirement, Occurrence

The daily requirement of vitamin B_{12} is shown in Table 6.2. The plasma concentration is normally 450 pg/mL.

The ability of vitamin B_{12} to promote growth alone or together with antibiotics, for example in young chickens, suckling pigs and young hogs, is of particular importance. This effect appears to be due to the influence of the vitamin on protein metabolism, and it is used in animal feeding. The increase in feed utilization is exceptional with underdeveloped young animals. Vitamin B_{12} is of importance also in poultry operations (enhanced egg laying and chick hatching). The use of vitamin B_{12} in animal feed vitamin

FOOD CHEMISTRY

(Formula 6.21)

fortification is obviously well justified.

The liver, kidney, spleen, thymus glands and muscle tissue are abundant sources of vitamin B_{12} (Table 6.4). Consumption of internal organs (variety meats) of animals is one method of alleviating vitamin B_{12} deficiency symptoms in humans.

6.3.9.2 Stability, Degradation

The stability of vitamin B_{12} is very dependent on a number of conditions. It is fairly stable at pH 4-6, even at high temperatures. In alkaline media or in the presence of reducing agents, such as ascorbic acid or SO_2, the vitamin is destroyed to a greater extent.

Glossary

biotin	生物素
cholecalciferol	维生素 D_3（胆钙化素）
chromoproteins	色蛋白
cyanocobalamin	维生素 B_{12}（氰钴胺）
degradation	降解
fat-Soluble Vitamins	脂溶性维生素
maillard-type browning reactions	美拉德褐变反应
naphthoquinone	萘醌
nicotinamide	烟酰胺

pyridoxine	维生素 B₆（吡哆醇）
retinol	维生素 A（视黄醇）
riboflavin	核黄素
thiamine	硫胺素
water-Soluble Vitamins	水溶性维生素

Chapter 7 Enzymes

7.1 Introduction

Enzymes are proteins with powerful catalytic activity. They are synthesized by biological cells and in all organisms, and involved in biochemical reactions related to the metabolism without itself being altered in the process. Therefore, enzyme-catalyzed reactions also proceed in many foods and thus enhance or deteriorate the food quality. Relevant to this phenomenon in the ripening of fruits and vegetables, the aging of meat and dairy products, and the processing steps involved in the making of dough from wheat or rye flours and the production of alcoholic beverages by the fermentation technology. Enzyme inactivation or changes in the distribution patterns of enzymes in subcellular particles of a tissue can occur during storages and thermal treatment in food. Since such changes are easily detected by analytical means, enzymes often act as suitable indicators for revealing food treatments. For example, pasteurization of milk, beer and honey, and differentiations between fresh and deep frozen meat or fish.

Enzyme properties attracted food chemists since enzymes are available in increasing numbers for the enzymatic food analysis or for utilization in the industrial food processing. Details of enzymes which play an important role in food science are restricted in this chapter to only those enzyme properties which are able to provide an insight into the build-up or functionality of enzymes or can contribute to the understanding of enzyme utilization in food analysis or food processing and storage.

7.2 General Remarks

7.2.1 Catalysis

Let us consider the catalysis of an exergonic reaction (Formula 7.1) with a most frequently occurring case in which the reaction does not proceed spontaneously. Reactant

A is metastable, since the activation energy, EA, required to reach the activated transition state in which chemical bonds are formed or cleaved in order to yield product P, is exceptionally high (Fig. 7.1).

$$A \underset{k_{-1}}{\overset{k_1}{\rightleftharpoons}} P \qquad \text{(Formula 7.1)}$$

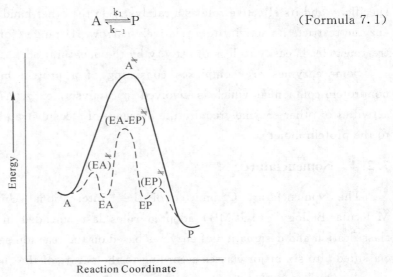

Fig. 7.1 Energy profile of an exergonic reaction A→P; — without and with catalyst E

The reaction is accelerated by the addition of a suitable catalyst. It transforms reactant A into intermediary products (EA and EP in Fig. 7.1), the transition state of which is at a lower energy level than the transition state of a noncatalyzed reaction (A≠ in Fig. 7.1). The molecules of species A contain enough energy to combine with the catalyst and, thus, to attain the "activated state" and to form or break the covalent bond that is necessary to give the intermediary product which is then released as the product P along with the free, unchanged catalyst. The reaction rate constants, k_{+1} and k_{-1}, are therefore increased in the presence of a catalyst. However, the equilibirum constant of the reaction, i. e., the ratio $k_{1+}/k_{-1}=K$, is not altered.

7.2.2 Specificity

Except for an enzyme's ability to substantially improve reaction rates, there is an unique enzyme property related to its high specificity for both the compound to be converted (substrate specificity) and for the type of reaction to be catalysed (reaction specificity).

The activities of allosteric enzymes are affected by specific regulators or effectors. Thus, the activities of such enzymes show an additional regulatory specificity.

7.2.3 Structures

Enzymes are globular proteins with greatly different particle sizes. The protein structure is determined by its amino acid sequences and conformation, both secondary and tertiary, derived from this sequence. Larger enzyme molecules often consist of two or more peptide chains (subunits or protomers) arranged into a specified quaternary structure. Three dimensional shape of the enzyme molecule is actually the reason for its specificity and its effective role as a catalyst. On the other hand, the protein nature of the enzyme restricts its activity to a relatively narrow pH range (for pH optima) and the heat treatment leads easily to loss of activity by the denaturation.

Some enzymes are complexes consisting of a protein moiety bound firmly to a nonprotein component which is involved in catalysis, e. g. a "prosthetic" group. The activities of other enzymes require the presence of a cosubstrate which is reversibly bound to the protein moiety.

7.2.4 Nomenclature

The Nomenclature Committee of the "International Union of Biochemistry and Molecular Biology" (IUBMB) adopted rules last amended in 1992 for the systematic classification and designation of enzymes based on the reaction specificity. All enzymes are classified into six major classes according to the nature of the chemical reaction catalyzed:

ⅰ) Oxidoreductases.

ⅱ) Transferases.

ⅲ) Hydrolases.

ⅳ) Lyases (cleave C—C, C—O, C—N, and other groups by the elimination, leaving double bonds, or conversely adding groups to double bonds).

ⅴ) Isomerases (involved in the catalysis of isomerizations within one molecule).

ⅵ) Ligases (involved in the biosynthesis of a compound with the simultaneous hydrolysis of a pyrophosphate bond in ATP or a similar triphosphate).

Each class is then subdivided into subclasses which more specifically denote the type of the reaction, e. g. by naming the electron donor of an oxidation-reduction reaction or by naming the functional group carried over by a transferase or cleaved by a hydrolase enzyme.

Each subclass is further divided into sub-subclasses. For example, sub-subclasses of oxidoreductases are denoted by naming the acceptor which accepts the electron from its respective donor.

Each enzyme is classified by adopting this system. An example will be analyzed. The enzyme ascorbic acid oxidase catalyzes the following reaction:

$$\text{L-Ascorbic acid} + \frac{1}{2}O_2 \rightleftharpoons \text{L-Dehydroascorbic acid} + H_2O \qquad (\text{Eq. 7.2})$$

Hence, its systematic name is L-ascorbate: oxygen oxidoreductase, and its systematic number is E. C. 1.1.10.3.3 (cf. Formula 7.3). The systematic names are often quite long. Therefore, the short, trivial names along with the systematic numbers are more convenient for the enzyme designation. Since enzymes of different biological origins often vary in their properties, and, when known, the subcellular fraction used for the isolation are specified in addition to the name of the enzyme preparation; for example, "ascorbate oxidase (E. C. 1.1.10.3.3) from cucumber". When known, the subcellular fraction of origin (cytoplasmic, mitochondrial or peroxisomal) is also specified.

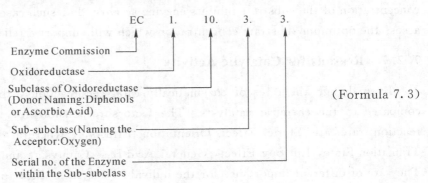

(Formula 7.3)

7.2.5 Activity Units

The catalytic activity of enzymes is exhibited only under specific conditions, such as pH, ionic strength, buffer type, presence of cofactors and suitable temperature. Therefore, the rate of substrate conversion or product formation can be measured in a test system designed to follow the enzyme activity. The International System of Units (SI) designation is $mol \cdot s^{-1}$ and its recommended designation is the "katal" (kat $*$). Decimal units are formed in the usual way, e. g.:

$$\mu kat = 10^{-6} kat = \mu mol \cdot s^{-1} \qquad (Eq.\ 7.4)$$

7.2.6 Determination of Enzyme Activity

In the foreword of this chapter it was emphasized that enzymes were suitable indicators for identifying food with. However, the determination of enzyme activity reaches far beyond this possibility: it is being increasingly used for the evaluation of the quality of raw food and for optimizing the parameters of particular food processes. In addition, the activities of enzyme preparations have to be controlled prior to use in processing or in enzymatic food analysis.

The measuring method of the catalytic activity of an enzyme is the rate of the reaction catalyzed by the enzyme. The conditions of an enzyme activity assay are optimized with relation to: type and ionic strength of the buffer, pH, and concentrations of substrate, cosubstrate and activators used. The closely controlled assay conditions, including the temperature, are critical because, in contrast to the substrate analysis, the reliability of the results in this case often can not be verified by using a weighed standard sample.

Temperature is an especially critical parameter which strongly affects the enzyme assay. Temperature fluctuations significantly affect the reaction rate; e. g., a 1 ℃ increase in temperature leads to about a 10% increase in activity. Whenever possible, the incubation temperature should be maintained at 25 ℃.

It is often difficult to try to achieve often arise while trying to achieve the ideal condition, i. e., $[A_0] \gg K_m$: the substrate's solubility is limited; spectrophotometric readings become unreliable because of high light absorbance by the substrate; or the high concentration of the substrate inhibits enzyme activity. For such cases procedures exist to assess the optimum substrate concentration which will support a reliable activity assay.

7.2.7 Reasons for Catalytic Activity

Even though the rates of enzymatically catalyzed reactions, they are much higher compared to the chemical catalysts. The factors responsible for the high increase in reaction rate are Steric Effects-Orientation Effects, Structural Complementarity to Transition State, Entropy Effect, General Acid-Base Catalysis, and Covalent Catalysis. They are of different importance for the individual enzymes. Structural Complementarity to Transition State is assumed that the active conformation of the enzyme matches the transition state of the reaction. This is supported by affinity studies which shows that a compound with a structure analogous to the transition state of the reaction ("transition state analogs") is bound better than the substrate. Hydroxamic acid, for example, is such a transition state analog which inhibits the reaction of triosephosphate isomerase (Fig. 7.2). The comparison between the Michaelis constant and the inhibitor constant shows that the inhibitor has a 30 times higher affinity to the active site than the substrate.

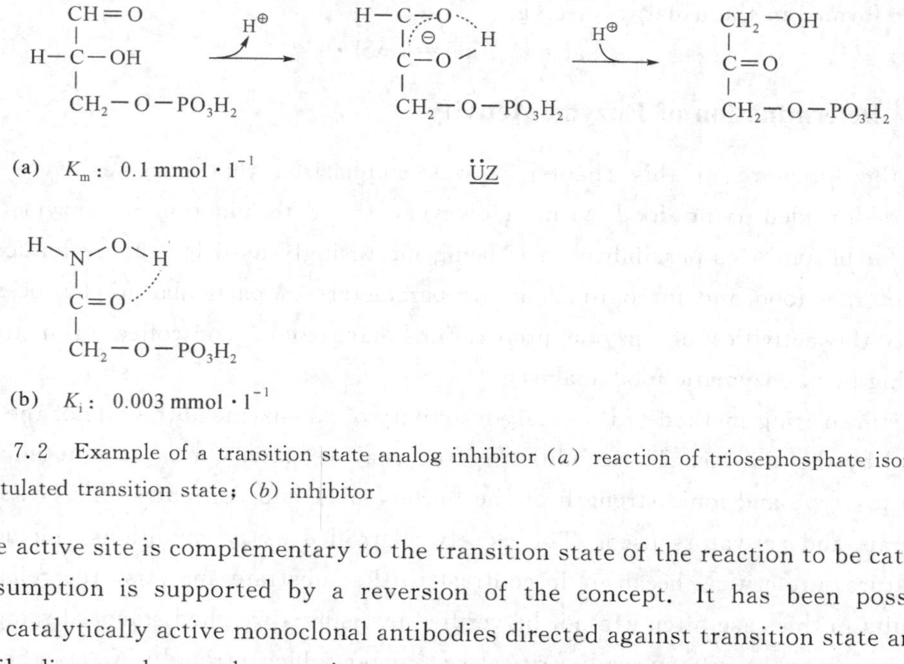

(a) K_m: 0.1 mmol·l^{-1} ÜZ

(b) K_i: 0.003 mmol·l^{-1}

Fig. 7.2 Example of a transition state analog inhibitor (a) reaction of triosephosphate isomerase, TT: postulated transition state; (b) inhibitor

The active site is complementary to the transition state of the reaction to be catalyzed. This assumption is supported by a reversion of the concept. It has been possible to produce catalytically active monoclonal antibodies directed against transition state analogs. The antibodies accelerate the reaction approximating the transition state of the analog. However, their catalytic activity is weaker compared to enzymes because only the environment of the antibody which is complementary to the transistion state causes the acceleration of the reaction.

Transition state analog inhibitors were used to show that in the binding the enzyme displaces the hydrate shell of the substrates. The reaction rate can be significantly increased by removing the hydrate shell between the participants.

The distortion of bonds and shifting of charges are also considered as important factors in catalytic veactions. The substrate's bonds will be strongly polarized by the enzyme, and thus highly reactive, through the precise positioning of an acid or base group or a metall ion (Lewis acid). These hypotheses are supported by investigations using suitable transition state analog inhibitors.

7.2.8 Kinetics of Enzyme-Catalyzed Reactions

Enzymes in food can be detected only indirectly by measuring their catalytic activity and, in this way, differentiated from other enzymes. This is the rationale for acquiring knowledge needed to analyze the parameters which influence or determine the rate of an enzyme-catalyzed reaction. The reaction rate depends on the concentrations of the components involved in the reaction. Here we mean primarily the substrate and the enzyme. Also, the reaction can be influenced by the presence of activators and inhibitors. Finally, the pH, the ionic strength of the reaction medium, the dielectric constant of the solvent (usually water) and the temperature exert an effect.

Effect of Substrate Concentration

The effect of substrate concentration on velocity of product formation is shown in Fig. 7.3 for one-substrate reactions when the reactions follow Michaelis-Menten kinetics. Determination of V_{max} and K_m from data plotted as in Fig. 7.3 is difficult because V_{max} is achieved only when $[S]_0 > 100 K_m$ (zero order with respect to substrate concentration). The substrate may be insoluble and/or expensive at the concentrations needed, the substrate may inhibit the reaction at high substrate concentrations (Fig. 7.3), or it may activate the reaction at high substrate concentrations (Fig. 7.3). For these and other reasons, Lineweaver and Burk in 1934 showed that the Michaelis-Menten equation (Eq. 7.5) can be transformed from a right hyperbola to a straight line (Eq. 7.6) by taking the reciprocal to give where $1/v_0 = y$, $1/[S]0 = x$, K_m/V_{max} is the slope and $1/V_{max}$ is the y-intercept of the straight-line plot, and $-1/K_m$ is the x-intercept (Fig. 7.5). Therefore, V_{max} and K_m can be readily determined, using all the experimental data. For best results the $[S]_0$ used to obtain data for Fig. 7.5 should range from $0.2 K_m$ to $5 K_m$, if possible. The Lineweaver-Burk method for calculating V_{max} and K_m is, by far, the most used. Other linear transforms include the methods of Augustinsson (Eq. 7.7) and of Eadie-Hofstee (Eq. 7.8).

$$v_0 = \frac{V_{max}[S]_0}{K_m + [S]_0} = \frac{k_2[E]_0[S]_0}{K_m + [S]_0} \qquad \text{(Eq. 7.5)}$$

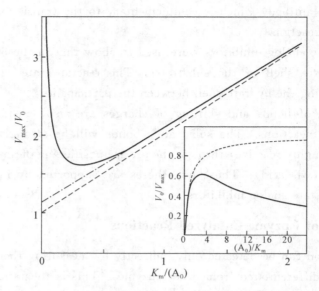

Fig. 7.3 Effect of substrate inhibition on the initial velocity of an enzyme-catalyzed reaction. The dashed line shows the normal curve in absence of inhibition, and the solid line shows inhibition by a second substrate molecule for which the dissociation constant, $K_{sr'}$ is 10 K_m. The solid line is calculated according to the equation $v_0 = V_{max}/\{[1+K_m/(A_0)]+[A_0/K_{sr}']\}$. E. A_2 does not form product. Large graph plotted by Line weaver-Burk method; the insert is a Michaelis-Menten plot. (Whitaker, 1994).

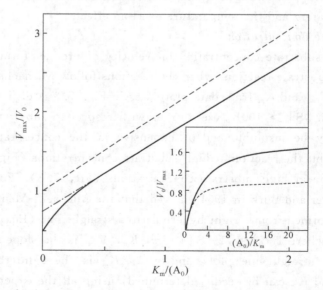

Fig. 7.4 Effect of substrate activation on the initial velocity of an enzyme-catalyzed reaction. The dashed line shows a normal reaction in absence of activation, and the solid line shows a reaction in the presence of activation. The solid line is calculated on the the assumptions that $K_{sr'}$ for the second substrate molecule is 10 K_m, that V_{max} is doubled when all the enzyme is saturated with a second substrate molecule (i.e., E. S_2 goes to product twice as fast as E. S), and that the second substrate molecule does not form the product. The large graph is plotted by the Line weaver-Burk method; the insert is a Michaelis-Menten plot. (Whitaker, 1994)

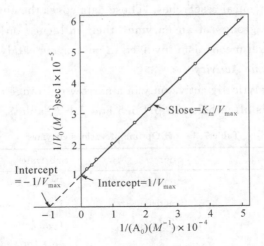

Fig. 7.5 Plot of substrate-velocity data according to the Lineweaver-Burk method as shown in Eq. 7.6. (Whitaker, 1994)

$$\frac{1}{v_0} = \frac{K_m}{V_{max}[S]_0} + \frac{1}{V_{max}} \quad \text{(Eq. 7.6)}$$

$$v_0 = V_{max} - \frac{v_0}{[S]_0} K_m \quad \text{(Eq. 7.7)}$$

$$\frac{[S]_0}{v_0} = \frac{K_m}{V_{max}} + \frac{[S]_0}{V_{max}} \quad \text{(Eq. 7.8)}$$

Effect of Inhibitors

The catalytic activity of an enzyme, in addition to substrate concentration, is affected by the type and concentration of inhibitors, i. e., compounds which decrease the rate of catalysis, and activators, which have the opposite effect. Metal ions and compounds which are active as prosthetic groups or which provide stabilization of the enzyme's conformation or of the enzyme-substrate complex are activators. The effect of inhibitors will be discussed in more details in this section.

Inhibitors are found among food constituents. Proteins which specifically inhibit the activity of certain peptidases, amylases or β-fructofuranosidase are examples. Furthermore, food contains substances which nonselectively inhibit a wide spectrum of enzymes. Phenolic constituents of food and mustard oil are included in this group. In addition, food might be contaminated with pesticides, heavy metal ions and other chemicals from a polluted environment which can become inhibitors under some circumstances. These possibilities should be taken into account when enzymatic food analysis is performed.

Food is usually heat treated to inhibit undesired enzymatic reactions. As a rule, no inhibitors are used in food processings. except an example that, the additionof SO_2 is used to inhibit the activity of phenolase.

Much data concerning the mechanism of action of enzyme inhibitors have been compiled in recent biochemical researches. These data cover the elucidation of the effect of inhibitors on functional groups of an enzyme, their influence on the active site and the clarification of the general mechanism involved in an enzyme catalyzed reaction.

Effects of pH on Enzyme Activity

Each enzyme is catalytically active only in a narrow pH range and, as a rule, each has an optimum pH which is often between pH 5.5 and 7.5 (Table 7.1).

Table 7.1 pH Optima of various enzymes

Enzyme	Source	Substrate	pH Optimum
Pepsin	Stomach	Proteln	2.0
Chymotrypsin	Pancreas	Protein	7.8
Papain	Tropical plants	Protein	7-8
Lipase	Microorganlsms	Olive oil	5-8
α-Glucosidase (maltase)	Microorganisms	Maltose	6.6
β-Amylase	Malt	Starch	5.2
β-Fruclofuranosidase (invertase)	Tomato	Saccharose	4.5
Pectin lyase	Microorganisms	Pectic acid	9.0-9.2
Xanthine oxidase	Milk	Xanthine	8.3
Lipoxygenase, type I [a]	Soyean	Linoleic acid	9.0
Lipoxygenase, type II [b]	Soybem	Linoleic acld	6.5

The optimum pH is affected by the type and ionic strength of the buffer used in the assay. The reasons for the sensitivity of the enzyme to changes in pH are two-fold: (i) Sensitivity is associated with a change in protein structure leading to the irreversible denaturation, (ii) The catalytic activity is dependent on the quantity of electrostatic charges on the enzyme's active site generated by the prototropic groups of the enzyme.

In addition, the ionization of dissociable substrates as affected by pH can be important to the reaction rate. However, such effects should be determined separately. Here, only the influences mentioned under b will be considered with some simplifications.

Influence of Temperature

Experimental data on the effect of temperature on velocities of enzyme-catalyzed reactions can be just as confusing and uninterpretable as the effect of pH, unless the experiments are designed properly. Temperature affects not only the velocity of catalysis of

$$E \cdot S \xrightarrow{k_2} E + P$$

but also stability of the enzyme; the equilibria of all association/dissociation reactions [ionization of buffer, substrate, product and cofactors (if any)]; association/disassociation of enzyme-substrate complex; reversible enzyme reactions solubility of substrates, especially gases; and ionization of prototropic groups in the active site of the enzyme and enzyme-substrate complex. To some extent, proper design of the experiments

for temperature effects is easier than that for pH effects.
$$S \rightleftharpoons P;$$

There are usually three reasons that temperature effects on enzymes are studied: (i) to determine stability of the enzyme; (ii) to determine the activation energy, E_a, of the enzyme-catalyzed reaction; and (iii) to determine the chemical nature of the essential prototropic groups in the active site of the enzyme. The design of the experiments is not difficult, except the buffer must always be made up to have the same pH at all temperatures used. This requires the buffer to be made up at the particular temperature. Otherwise, the pH of the buffer will be an uncontrolled rariability.

Influence of Pressure

The application of high pressures can inhibit the growth of microorganisms and the activity of enzymes. This could protect sensitive nutrients and aroma substances in foods. Some products preserved in this gentle way are now marketable. Microorganisms are relatively sensitive to high pressure because their growth is inhibited at pressures of 300-600 MPa and lower pH values increase this effect. However, bacterial spores withstand pressures above 1,200 MPa.

In contrast to thermal treatment, high pressure does not attack the primary structure of proteins at the room temperature. Only H-bridges, ionic bonds and hydrophobic interactions are disrupted. Quaternary structures are dissociated into subunits by comparatively low pressures (<150 MPa). Higher pressures (>1,200 MPa) change the tertiary structure and very high pressures disrupt the H-bridges which stabilize the secondary structure. The hydration of proteins is also changed by the high pressure because water molecules are pressed into cavities which can exist in the hydrophobic interior of proteins. In general, proteins are irreversibly denatured at the room temperature by the application of pressures above 300 MPa while lower pressures only cause reversible changes in the protein structure.

In the case of enzymes, even slight changes in the steric arrangement and mobility of the amino acid residues which participate in catalysis can resultin loss of activity. Nevertheless, a relatively high pressure is often required to inhibit enzyme activity. But the pressure required can be reduced by increasing the temperature, as shown in Fig. 7.6 for α-amylase. While a pressure of 550 MPa is required at 25 ℃ to inactivate the enzyme with a rate constant (first order reaction) of $k=0.01 min^{-1}$, a pressure of 340 MPa is only required at 50 ℃.

It is remarkable that enzymes can be activated by changes in the conformation of the polypeptide chain, which are initiated especially by low pressures around 100 MPa. In the application of the pressure technique for stable food processing, intact tissue, and not isolated enzymes, is exposed to high pressures. Thus, the enzyme activity can increase rather than decrease when cells or membranes are disintegrated with the release of enzyme and/or substrate.

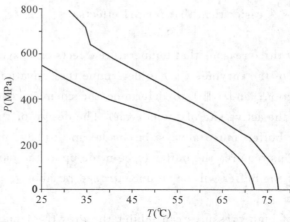

Fig. 7.6 Pressure-temperature diagram for the inactivation kinetics of α-amylase from Bacillus subtilis at pH 8.6 (Ludikhuyze *et al.*, 1997). Range of the rate constants: $k=0.01$ min^{-1} (lower line) to $k=0.07$ min^{-1} (upper line)

Influences of Water

Up to a certain extent, enzymes need to be hydrated in order to express activity. Hydration of lysozyme could be determined by the IR and NMR spectroscopy. As can be seen in Table 7.2, first the charged polar groups of the side chains hydrate, followed by the uncharged ones. Enzymatic activity starts at a water content of 0.2 g/g protein, which means even before a monomolecular layer of the polar groups with water has taken place. Increase in hydration leading to a monomolecular layer of the whole available enzyme surface at 0.4 g/g protein increases the activity to a limiting value reached at a water content of 0.9 g/g protein. Then the diffusion of the substrate to the enzyme's active site seems to be completely guaranteed.

Table 7.2 **Hydration of Lysozyme**

$\dfrac{\text{g Water}}{\text{g Protein}}$	Hydration sequence	Molecular changes
0.0	Charged groups Uncharged, polar groups (formation of clusters)	Relocation of protons New protons of disulfide bonds
0.1	Saturation of COOH groups Saturation of polar groups in side chains	Change in conformation
0.2	Peptide-NH	Start of enzymatic activity
0.3	Peptide-CO Monomolecular hydration of polar groups Apolar side chains	
0.4	Complete enzyme hydration	

For storage of food, it is mandatory to inhibit enzymatic activity completely if the storage temperature is below the phase transition temperature T_g or $T_{g'}$ (cf. 0.3.3). With help of a model system containing glucose oxidase, glucose and water as well as sucrose and maltodextrin (10 DE) for adjustment of $T_{g'}$ values in the range of 9.5 to 32 ℃, it was found that glucose was enzymatically oxidized only in such samples that were stored for two months above the $T_{g'}$ value instead of those kept at storage temperatures below $T_{g'}$.

7.3 Enzymes in Food

7.3.1 Amylases

Amylases, the enzymes that hydrolyze starches, are found not only in animals, but also in higher plants and microorganisms. Therefore, it is not surprising that some starch degradation occurs during the maturation, storage, and processing of our foods. Since starch contributes to viscosity and texture of foods in a major way, its hydrolysis during the storage and processing is a matter of importance. There are three major types of amylases: α-amylases, β-amylases, and glucoamylases. They act primarily on both starch and glycogen.

Amylases are either produced by the bacterias or yeasts and they belong to the components of malt preparations. The high temperature-resistant bacterial amylases, particularly those of *Bac. licheniformis* (Fig. 7.7) are of interest for the hydrolysis of the corn starch (gelatinization at 105-110℃). The hydrolysis rate of these enzymes can be enhanced further by adding Ca^{2+} ions. α-Amylases added to the wort in the beer production process accelerate starch degradation. These enzymes are also used in the baking industry.

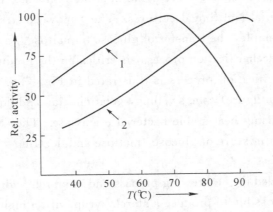

Fig. 7.7 The activity of α-amylase as influenced by temperatures. 1 α-amylase from Bacillus subtilis, 2 from Bacillus licheniformis.

α-Amylases

The α-amylases, found in all organisms, hydrolyze the interior α-1,4-glucosidic bonds of starch (both amylose and amylopectin), glycogen, and cyclodextrins with retention of the α-configuration of the anomeric carbon. Since the enzyme is endo-splitting, it has a great effect on the viscosity of starch-based foods, such as puddings, cream sauces, etc. The salivary and pancreatic α-amylases are very important in digesting starch in foods. Some microorganisms contain high levels of α-amylases. Some of the microbial α-amylases have high inactivation temperatures, and, if not activated, they can have a drastic undesirable effect on the stability of starch-based foods.

β-Amylases

β-Amylases, found in higher plants, hydrolyze the α-1,4-glucosidic bonds of starch at the nonreducing end to give β-maltose. Since they are exo-splitting enzymes, many bonds must be hydrolyzed before an appreciable effect on viscosity of starch paste is observed. Amylose can be hydrolyzed to 100% maltose by β-amylase, while β-amylase cannot continue beyond the first α-1,6-glycosidic bond encountered in amylopectin. Therefore, amylopectin is hydrolyzed only to a limited extent by β-amylase alone. "Maltose" syrups, of about DP 10, are great of importance in the food industry. β-Amylase, along with α-amylase, plays an important role in brewing, since the maltose can be rapidly converted to glucose by the yeast maltase. β-Amylase is a sulfhydryl enzyme and can be inhibited by a number of sulfhydryl group reagents, unlike α-amylase and glucoamylase. In malt, β-amylase is often covalently linked, via disulfide bonds, to other sulfhydryl groups; therefore, malt should be treated with a sulfhydryl compound, such as cysteine, to increase its activity in malt.

Glucoamylase

Glucoamylase cleaves β-D-glucose units from the non-reducing end of an 1,4-α-D-glucan. The α-1,6-branching bond present in amylopectin is cleaved at a rate about 30 times slower than the α-1,4-linkages occurring in straight chains. The enzyme preparation is produced from bacterial and fungal cultures. The removal of transglucosidase enzymes which catalyze, for example, the transfer of glucose to maltose, thus lowering the yield of glucose in the starch saccharification process, is critical in the production of glucoamylase.

The starch saccharification process is illustrated in Fig. 7.8. In a purely enzymatic process (left side of the figure), the swelling, gelatinization and liquefaction of starch can occur in a single step using heat-stable bacterial α-amylase. The amylase is used to yield starch syrup which is a mixture of glucose, maltose and dextrin.

Isoamylase

Isoamylase is utilized in the brewing process and in starch hydrolysis. In combination with β-amylase, it is possible to produce a starch syrup with a high maltose content.

Other Amylases

Other starch-splitting enzymes also exist (Table 7.3).

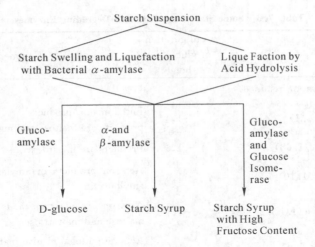

Fig. 7.8 Enzymatic starch degradation

Application

Amylase enzymes are used in bread making and to break down complex sugars such as starch (found in flour) into simple sugars. Yeast then feeds on these simple sugars and converts them into alcohol and CO_2. It confers flavour and causes the bread to rise. While amylase enzymes are found naturally in yeast cells, it takes time for the yeast to produce enough enzymes to break down significant quantities of starch in the bread. This is the reason for doughs such as sour dough forment for long time. Modern bread making techniques have included amylase enzymes (often in the form of malted barley) into bread improver thereby making the bread making process faster and more practical for the commercial use.

When used as a food additive, amylase has E number E1100, and it may be derived from swine pancreas or mould mushroom. Bacilliary amylase is also used in clothing and dishwasher detergents to dissolve starches from fabrics and dishes.

Workers in factories that work with amylase for any of the above uses are at increased risk of occupational asthma. 5%-9% of bakers have a positive skin test, and a fourth to a third of bakers with breathing problems are hypersensitive to amylase.

An inhibitor of alpha-amylase called phaseolamin has been tested as a potential diet aid.

Blood serum amylase is usually measured for purposes of the medical diagnosis. A normal concentration is in the range 21-101 U/L. A higher than normal concentrations may reflect one of several medical conditions, including acute inflammation of the pancreas (concurrently with the more specific lipase), perforated peptic ulcer, torsion of an ovarian cyst, strangulation ileus, macroamylasemia and mumps. Amylase may also be measured in other body fluids, including urine and peritoneal fluid.

Table 7.3 Some Starch-and Glycogen-Degrading Enzymes

Type	Configuration of glucosidie bond	Comments
*Endes*plitting(configuration retained)		
-Amylase(EC 3.2.1.1)	−1,4	Initial major products are dextrins; final major products are maltose and maltotriose
Isoarnylasc(EC 3.2.1.68)	−1,6	Products are lincardextrins
Isomaltasc(EC 3.2.1.10)	−1,6	Acts on products of amylase, hydrolysis of the amylopectin
Cyclomaltodextrinase(EC 3.2.1.54)	−1,4	Acts on cyclodextrins and lincardextrins to give maltose and maltotriose
Pullulanase(EC 3.2.1.41)	−1,6	Acts on pullulan to give maltotriose and on starch to give lineardextrins
Isopullulanase(EC 3.2.1.57)	−1,4	Acts on pullulan to give isopanose and on starch to give unknown products
Ncopullulanase	−1,4	Acts on pullulan to give panose and on starch to give maltose
Amylopullulanase	−1,6	Acts on pululan to give maltotroise
	−1,4	Acts on starch to give DP 2−4 products
Amylopectin 6-glucanohydrolase (EC 3.2.1.41)	−1,6	Acts only on amylopectin to hydrolyzs 1,6-glucosidic linkages
*Exo*splitting(nonreducing end)		
-Amylase(EC 3.2.1.2)	−1,4	Productis -maltose
-Amylase	−1,4	Productis -maltose; there are specificre -amylases that producc maltotriose, mallotertraose, maltopentaose, and maltohexaose, with retention of configuration
Glucoamylase(EC 3.2.1.3)	−1,6	-Glucose is produced
-Glucosidase(EC 3.2.1.20)	−1,4	-Glucose is produced; there are a number of -glucosidases
Transferase		
Cyclomaltodextrin gluc anotransfcrase (EC 2.4.1.19)	−1,4	-and -Cyclode xtrins formed from starch with 6−12 glucose units

In molecular biology, the presence of amylase can serve as an additional method of selecting for successful integration of a reporter construct in addition to the antibiotic resistance. As reporter genes are flanked by homologous regions of the structural gene for amylase, successful integration will disrupt the amylase gene and prevent the starch degradation, which is readily detectable through the iodine staining.

7.3.2 Pectic Enzymes

The pectic enzymes include a diverse group of enzymes. Pectic acid which is liberated by pectin methylesterases flocculates in the presence of Ca^{2+} ions. This reaction is responsible for the undesired "cloud" flocculation in citrus juices. After the thermal inactivation of the enzyme at about 90 ℃, this reaction is not observable. However, such treatment brings about the deterioration of the aroma of the juice. Investigations of the pectin esterase of the orange peel have shown that the enzyme activity is affected by competitive inhibitors: oligogalacturonic acid and pectic acid (cf. Fig. 7.9). Therefore, the increase in turbidity of citrus juice can be prevented by the addition of such compounds.

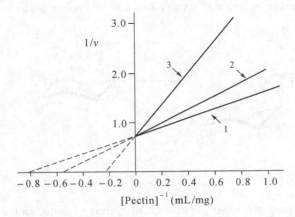

Fig. 7.9 Pectin esterase (orange) activity as affected by inhibitors (according to Termote, 1977). 1: Without inhibitor; 2: hepta-and octagalacturonic acids; 3: pectic acid

Pectin Methylesterase

Pectin methylesterase (pectin pectylhydrolase, EC 3.1.1.11) hydrolyzes the methyl ester bond of pectin to give pectic acid and methanol (Eq. 7.9).

(Eq. 7.9)

Polygalacturonase

Polygalacturonase (poly-a-1,4-galacturonide glycano-hydrolase, EC 3.2.1.15) hydrolyzes the a-1,4-glycosidic bond between the anhydrogalacturonic acid units (Eq. 7.10).

FOOD CHEMISTRY

(Eq. 7.10)

Pectate lyase

The pectate lyases [poly(1, 4-aD-galacturonide) lyase, EC 4.2.2.2] split the glycosidic bond of both pectin and pectic acid, not with water, but by β-elimination (Eq. 7.11). They are found in microorganisms, but not in higher plants.

(Eq. 7.11)

Application

Pectinolytic enzymes are used for the clarification of fruit and vegetable juices. The mechanism of the clarification is as follows: the core of the turbidity causes particles consisting of carbohydrates and proteins (35%). The prototropic groups of these proteins have a positive charge at the pH of fruit juice (3.5). Negatively charged pectin molecules form the outer shell of the particle. Partial pectinolysis exposes the positive core. Aggregation of the polycations and the polyanions then follows, resulting in the flocculation. The clarification of juice by gelatin (at pH 3.5 the gelatin is positively charged) and the inhibition of the clarification by alginates which are polyanions at pH 3.5 support this suggested model.

In addition, pectinolytic enzymes play an important role in food processings, increasing the yield of fruit, vegetable juices and oil from olive fruits.

Table 7.4 Applications of pectic enzymes

Food	Purpose or action
Fruit juices	Improving yield of press juices, preventing cloudiness, and improving concentratin processes
Olives	Extracting oil
Wines	Clarification

7.3.3 Cellulases and Hemicellulases

The baking quality of rye flour and the shelf life of rye bread can be improved by partial hydrolysis of the rye pentosans. Technical pentosanase preparations are mixtures of β-glycosidases (1,3-and 1,4-β-D-xylanases, etc.).

Solubilization of plant constituents by soaking in an enzyme preparation (maceration) is a mild and sparing process. Such preparations usually contain exo-and endo-cellulases, α- and β-mannosidases and pectolytic enzymes. Examples of the utilization are: production of fruit and vegetable purées (mashed products), disintegration of tea leaves, or production of dehydrated mashed potatoes. Some of these enzymes are used to prevent mechanical damage to cell walls during mashing and, thus, to prevent the excessive leaching of gelatinized starch from the cells, which would make the purée too sticky.

Glycosidases (cellulases and amylases from *Aspergillus niger*) in combination with proteinases are recommended for removal of shells from shrimp. The shells are loosened and then washed off in a stream of water.

Table 7.5 Applications of pectic Cellulases and Hemicellulases

Food	Purpose or action
Chocolate/cocoa	Hydrolytic activity during the fermentation of cocoa
Coffee	Hydrolysis of gelatinous coating during the fermentation of beans
Fruits	Softening

7.3.4 Protease

Texture of food products can be changed by hydrolysis of proteins by endogenous and exogenous proteases. For examlpe, Gelatin will not gel when raw pineapples is added, because the pineapples contains bromelain, a protease. Chymosin causes milk to gel, as a for the reasin that its hydrolysis of a single peptide bond between Phe105-Met106 in k-casein. This specific hydrolysis of k-casein destabilizes the casein micelle, causing it to aggregate to form curds (cottage cheese). Action of intentionally added microbial proteases during aging of brick cheeses assists in development of flavors (flavors in Cheddar cheese vs. blue cheese, for example). Protease activity on the gluten proteins of wheat bread doughs during rising is important not only in the mixing characteristics and energy requirements but also in the quality of the baked breads.

Serine Proteases

Serine proteases or serine endopeptidases (newer name) are proteases (enzymes that cut peptide bonds in proteins) in which one of the amino acids in the active site of the enzyme is serine.

They are found in both single-cell and complex organisms, in both cells with nuclei (eukaryotes) and without nuclei (prokaryotes). Serine proteases are grouped into clans that share structural similarities (homology) and are further subgrouped into families with similar sequences.

The major clans found in humans include the chymotrypsin-like, the subtilisin-like, the α/β hydrolase, and signal peptidase clans. More recently discovered have been the rhomboid family of intramembrane serine proteases.

In evolutionary history, serine proteases were originally digestive enzymes. In mammals, they evolved by gene duplications to serve functions in blood clotting, the immune system, and inflammation. Rhomboid family intramembrane serine proteases do not share common ancestry with the other major clans, and appear to have evolved by convergent evolution.

Serine proteases are paired with serine protease inhibitors, which turn off their activity when they are no longer needed.

The three serine proteases of the chymotrypsin-like clan that have been studied in greatest detail are chymotrypsin, trypsin, and elastase. All three enzymes are synthesized by the pancreatic acinar cells, secreted in the small intestine, and are responsible for catalyzing the hydrolysis of peptide bonds.

Aspartate Proteases

Aspartic proteases are a family of protease enzymes that use an aspartate residue for catalysis of their peptide substrates. In general, they have two highly-conserved aspartates in the active site and are optimally active at acidic pH. Nearly all known aspartyl proteases are inhibited by the pepstatin.

Aspartic endopeptidases EC 3.4.23. of vertebrate, fungal and retroviral origin have been characterized. More recently, aspartic endopeptidases associated with the processing of bacterial type 4 prepilin and archaean preflagellin have been described.

Eukaryotic aspartic proteases include pepsins, cathepsins, and renins. They have a two-domain structure, arising from ancestral duplication. Retroviral and retrotransposon proteases are much smaller and appear to be homologous to a single domain of the eukaryotic aspartyl proteases. Each domain contributes a catalytic Asp residue, with an extended active site cleft localized between the two lobes of the molecule. One lobe has probably evolved from the other through a gene duplication event in the distant past. In modern-day enzymes, although the three-dimensional structures are very similar, the amino acid sequences are more divergent, except for the catalytic site motif, which is very conserved. The presence and position of disulfide bridges are other conserved features of aspartic peptidases.

Metal-Containing Proteases

Metal-containing proteases constitute a family of enzymes from the group of proteases, classified by the nature of the most prominent functional group in their active site. These are proteolytic enzymes whose catalytic mechanism involves a metal. Most

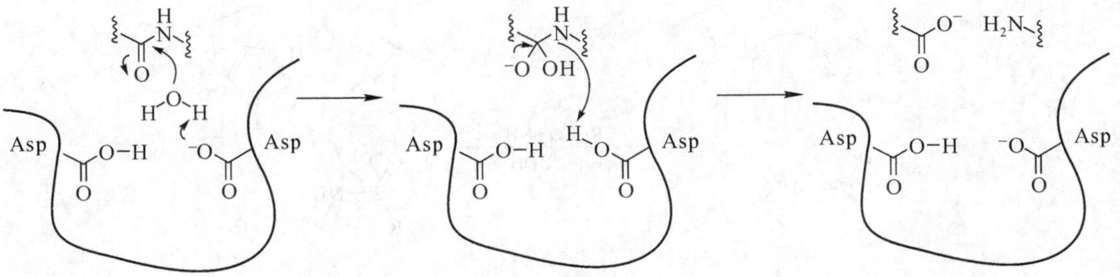

Fig. 7.10 Proposed mechanism of peptide cleavage by aspartyl proteases

metal-containing proteases are zinc-dependent, but some use cobalt. The metal ion is coordinated to the protein via three ligands. The ligands co-ordinating the metal ion can vary with histidine, glutamate, aspartate, lysine and arginine. The fourth coordination position is taken up by a labile water molecule. There are two subgroups of metal-containing proteases: exopeptidases: metalloexopeptidases; endopeptidases; metalloendopeptidases. Well known metalloendopeptidases include ADAM proteins and matrix metalloproteinases. Treatments with chelating agents such as EDTA leads to the complete inactivation. EDTA is a metal chelator which removes zinc, which is essential for activity. They are also inhibited by the chelator orthophenanthroline.

Cysteine Proteases

Cysteine proteases have a common catalytic mechanism that involves a nucleophilic cysteine thiol in a catalytic dyad. The first step is deprotonation of a thiol in the enzyme's active site by an adjacent amino acid with a basic side chain, usually a histidine residue. The next step is nucleophilic attack by the deprotonated cysteine's anionic sulfur on the substrate carbonyl carbon. In this step, a fragment of the substrate is released with an amine terminus, the histidine residue in the protease is restored to its deprotonated form, and a thioester intermediate linking the new carboxy-terminus of the substrate to the cysteine thiol is formed. Therefore they are also sometimes considered as thiol proteases. The thioester bond is subsequently hydrolyzed to generate a carboxylic acid moiety on the remaining substrate fragment, while regenerating the free enzyme.

Cysteine proteases are commonly encountered in fruits including papaya, pineapple, fig and kiwifruit. The proportion of protease tends to be higher when the fruit is unripe. In fact, dozens of latices of different plant families are known to contain cysteine proteases. Cysteine proteases are used as an ingredient in meat tenderizers.

Applications

The effect of proteases in the tenderization of meat is perhaps best known and economically most important. After death, muscle becomes rigid due to rigor mortis (caused by the extensive interaction of myosin and actin). Through action of endogenous proteases (Ca^{2+}-activated proteases, and perhaps cathepsins) on the myosin-action complex during the storage (7-21 days) the muscle becomes more tender and juicy.

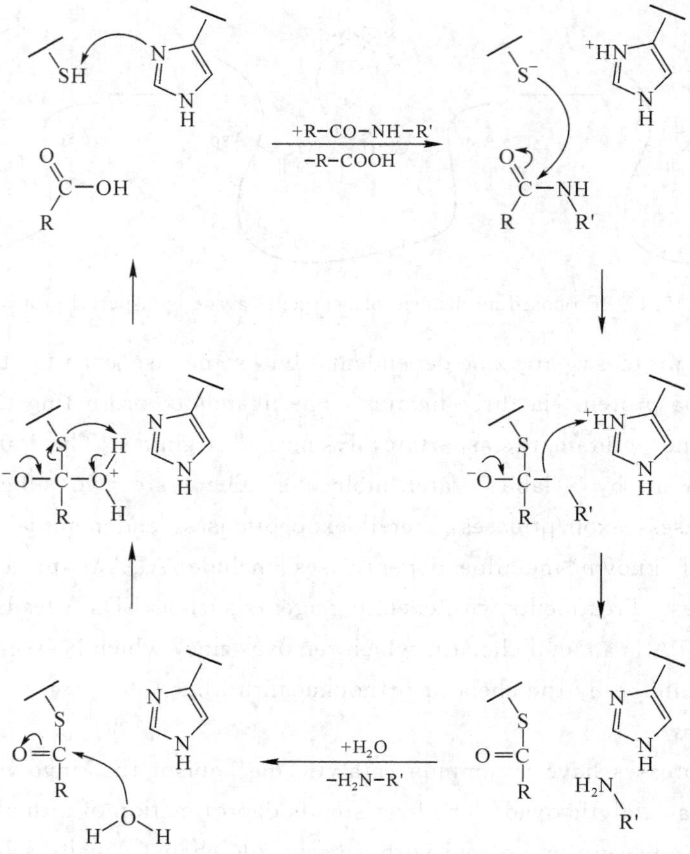

FOOD CHEMISTRY

Fig. 7.11 The reaction mechanism of the cysteine protease mediated cleavage of a peptide bond

Exogenous enzymes, such as papain and ficin, are added to some less choice meats to tenderize them, primarily due to the partial hydrolysis of elastin and collagen.

7.3.5 Lipases

Lipases perform essential roles in the digestion, transport and processing of dietary lipids (e.g. triglycerides, fats, oils) in most, if not all, living organisms. Genes encoding lipases are even present in certain viruses. Lipase from microbial sources (e.g. Candida lipolytica) is utilized for enhancement of aromas in the cheese making. Limited hydrolysis of milk has also been used in the production of chocolate milk. It enhances the "milk character" of the flavor.

Staling of bakery products is retarded by lipase, presumably through the release of mono-and diacylglycerols. The defatting of bones, which has to be carried out under mild conditions in the production of gelatin, is facilitated by using lipase-catalyzed hydrolysis.

7.4　Enzyme Utilization in the Food Industry

7.4.1　Starch

A considerable quantity of the sweeteners used throughout the world is derived from starch as opposed to cane or beet sugar. The enzymatic treatment of starch has become much more popular than acid hydrolysis.

The treatment of starch with enzymes results in a variety of sweet syrups used throughout the food and beverage industries. Three stages can be identified in the starch modification. Firstly, amylases liberate "maltodextrin" by the liquefaction process. Such maltodextrins are not very sweet since they contain dextrins and oligosaccharides.

The dextrins and oligosaccharides are further hydrolysed by enzymes such as pullulanase and glucoamylase in a process known as the saccharification. The complete saccharification converts all the limit dextrans to glucose, maltose and isomaltose. The resulting syrups are moderately sweet and are frequently modified further.

The treatment of glucose/maltose syrups with glucose isomerase converts a large proportion of the glucose to sweeter fructose. This isomerisation process is usually performed with immobilised glucose isomerase and results in syrups with approximately 50% fructose and 50% glucose. Such products are known as high fructose syrups and are frequently used as "sugar replacements" in the manufacture of foods and beverages.

7.4.2　Baking

Enzymes are rapidly becoming important to the baking industry. Enzyme supplements are used in baking to make consistently high-quality products by enabling better dough handling, providing anti-staling properties, and allowing control over crumb texture and color, taste, moisture, and volume.

Most bread is made of wheat, which has naturally occurring enzymes that, when water is added, modify the starch, protein, and fiber. The yeast that is added produces carbon dioxide from simple sugars, which makes the bread rise. However, the quality of wheat flour varies due to natural variations, the time of year it is picked, disparities in milling, and many other inconsistencies. Chemical supplements that were used in the baking industry are being replaced by enzymes. For instance, the enzyme oxidase can replace chemical oxidants, such as bromates, used to strengthen gluten.

Gluten is the viscous and elastic network found inside the bread that gives it its unique consistency and holds the carbon dioxide that makes the bread rise. The reason wheat is the primary grain used in baking, is that it's special gluten that is not found in other grain such as barley and rye, which produces denser and harder breads. Oftentimes, enzymes are added to other grains to weaken its gluten and produce biscuits, crackers, cookies and other crisp bread dough.

The standard procedure to make minor adjustments to the amylase levels of milled baking flours is to add small levels of fungal alpha amylase. Recent studies in many centers have shown that enzymes other than simple alpha amylase can contribute to the performance of the flour. It is also noted that application of non-amylase enzymes in addition to regular levels of amylase can make substantial improvements to flour behaviour. Furthermore the optimum adjustment required will vary according to the intended application of the flour and also as the cereal ages in store.

7.4.3 Brewing

Enzymes have proved to be useful for the brewing industry in many areas of beer productions. They can be added to the beer after its fermentation to induce faster maturations. Enzymes also work as filtration improvers, reducing the presence of viscous polysaccharides such as xylans and glucans. Enzymes are often used to remove carbohydrates in the production of light beer and induce chill proofing. The beer brewing involves the production of alcohol by allowing yeast to act on plant materials such as barley, maize, sorghum, and rice. Yeast converts simple sugars into alcohol and carbon dioxides. However, the sugars found in the plant materials are most often complex polysaccharides that yeast is unable to convert. The traditional method for breaking down these complex polysaccharides is called malting. This is the process whereby barley, for example, is allowed to partially germinate, producing enzymes that break down the complex polysaccharides into simple sugars that the yeast can utilize. However, the process of malting can be expensive and often difficult to control. By adding enzymes to unmalted barley the complex polysaccharides can be broken down to simple sugars and reduce or eliminate the costly and complicated process of malting.

7.4.4 Cheese (Flavour)

Lipase can be used to produce a good range of savoury and cheesy flavours. In addition a blend of proteases and peptidases are used for the production of cheese flavour with a protein rather than a fatty acid. This range can be tailored to give a cheese flavour.

7.4.5 Coffee & Tea

As far as it was developed, many products on market are used in coffee and tea processing. Bioclean TC has a blend of natural enzymes that gives manufacturing areas a 100 percent safe-detergent and alkali-free cleansing. Bioclean TC is dissolved in water and sprayed in areas that come in contact with leaf materials, such as on rolling tables, CTC rollers, conveyor belts and CFMs. The blend helps degrading biological material, which is then easily dislodged in cold water washing. Bioclean TC prevents bacterial growths by providing no food to sustain the bacteria. Since it is 100 percent natural, the blend doesn't impart any off-flavor or off-taste to the tea.

Specialty enzymes, a leading solution provider to the human healthcare, animal nutrition, and natural product processing industries, offers SPEEDOX. When added to the wash water or effluent stream in the coffee production process, SPEEDOX rapidly reduces organic loads in the wastewater that cause environmental pollution. SPEEDOX can reduce the pollution causing entities in the coffee plantation wastewater. SPEEDOX is a 100% natural, high quality and eco-friendly product. It improves the clarity of water and the water can be used for the irrigation. SPEEDOX also eliminates the foul odor near the effluent treatment tanks.

7.4.6 Dietetics

Many disorders and diseases are caused by an inability to digest adequately some components of different foods. This inability may only result in an increased level of flatulence, such as results from the soya bean consumption, or might be much more serious as in the case of milk and wheat intolerance. Digestive enzymes taken with food in the form of capsules or tablets can ease these problems and disorders. Dietetic enzymes include Pancreatin but the main emphasis is on microbial enzymes produced by fermentation. This offers safer and more the effective alternatives. With microbial enzymes it is possible to get types that work both in the stomach and intestines. This greatly enhances their effectiveness. Also it is possible to obtain synergistic blends of enzymes which are particularly efficient at breaking down complex molecules such as proteins. The other supplied novel enzymes such as glutaminase which greatly enhances the overall breakdown of wheat proteins. It is potential to help prevent the wheat intolerance and coeliac diseases.

7.4.7 Eggs

Enzymes are often popular in producing egg product and improving their qualities. For example, proteolytic enzymes are used to improve drying properties. The use of phospholipase in mayonnaise processing leads to a significant improvement on viscosity, with a rate of 30%-35%. The wasted salted egg white is potential to be used to produce diverse functional products like amino acids and short peptides.

7.4.8 Juice & Wine

Enzymes increase processing capacity and improve economy in the fruit juice and wine industries. The most commonly used enzymes in these industries are pectinases. Pectinases increase juice yields and accelerate juice clarification. They produce clear and stable single-strength juices, juice concentrates and wines, from not only core-fruits such as apples and pears, but also stone fruits, berries, grapes, citrus-fruits, tropical fruits and vegetables like carrots, beets and green peppers.

Fruit and Vegetable Juice

Specialty enzymes are pectinases that contain hemicellulase enzyme activities. These products not only increase juice yields, but also increase the color and health-promoting antioxidants in fruit and vegetable juices extracted by pressing or decanter the centrifuge. By reducing fruit and vegetable mash viscosity and improving the solid/liquid separation, they increase color extractions and juice volumes.

Pectinase and Amylase enzyme solutions speed up the filtration and prevent the storage or post-packaging haze formation by depectinizing and reducing starch in raw juices. Pectin and starch must be removed from freshly extracted juices prior to the filtration, fining and concentration. Pectinase and Amylase can reduce starch and pectin in raw fruits and juices, thus achieving clear and stable juices and juice concentrates.

Wine

Many biochemical reactions involved in the wine production are enzyme-catalyzed. They begin during the ripening and harvesting of grapes, and continue through the alcoholic and malolactic fermentation, clarification, and ageing. Winemakers often supplement naturally occurring grape enzymes with commercial enzymes to increase the production capacity of clear and stable wines with enhanced body, flavor and bouquet.

When added to grapes or musts, pectinase that contain hemicellulase enzyme products increase free-run juice volume and extraction of color, fermentable sugars and flavor components, as well as reduce pressing and fermentation time. These pectinase or pectinase containing hemicellulase products can increase free-run juice volumes by 20 to 30 percent and lower fermentation time by 30 to 50 percent by reducing grape-pectin viscosityies.

The rapid clarification and well-separated lees have a positive effect on finished wine flavor, texture and color. β-glucanase containing pectinase depectinize grape-musts during the fermentation or young wines prior to the fining and filtration. Grape musts and wines treated with β-glucanase containing pectinase are less viscous. They ferment, settle and mature more quickly.

A β-glucanase containing pectinase, is also used to degrade Botrytis-glucan. Wines made from overripe grapes infected with Botrytis cinerea mold are often difficult to clarify and filter due to high concentrations of Botrytis-produced glucan polysaccharides. The use of β-glucanase containing pectinase can accelarate the clarification and filtration.

Acid proteases clarify and stabilize some wines by reducing or removing naturally occurring and yeast synthesized, heat-labile proteins.

Proteins

Sequences of proteinases for the conversion of proteins to highly modified hydrolysates of small peptides with either very low bitterness and/or strong taste are available. These enzymes have been carefully selected to work together after extensive studies of their action of target proteins.

General proteinases also should be supported by uniquely potent exopeptidases acting to tidy up flavour characteristics and remove bitterness.

Glossary

allosteric	变构的
aspartate protease	天冬氨酸蛋白酶
bromat	溴酸盐
buffer type	缓冲型
catalysis	催化
covalent bond	共价键
cathepsin	组织蛋白酶
cellulase	纤维素酶
chelating agents	螯合物
chromatograph	色谱分析
chymosin	凝乳酶
chymotrypsin	胰凝乳蛋白酶
citrus	柑橘
cobalt	钴
collagen	胶原蛋白
cysteine	半胱氨酸
cytoplasmic	原生质的
decimal	十进制的
denaturation	变性
deprotonation	去质子化
diacylglycerol	甘油二酯
disulfide bridge	二硫键桥
dough	生面团
elastase	弹性蛋白酶
electrophoretic	电泳的
endo-cellulase	内切纤维素酶
endopeptidase	肽链内切酶
endo-splitting	内切
equilibrium	平衡

FOOD CHEMISTRY

esterase	酯酶
exergonic	放能的
flocculates	絮凝剂
glucoamylase	葡糖淀粉酶
glutamate dehydrogenase	谷氨酸脱氢酶
glutamate-oxalacetate transaminase	谷草转氨酶
glycosidases	内糖苷酶
hemicellulases	半纤维素酶
hemicelluloses	半纤维素类
histidine	组氨酸
homologous	同源
hydrolase	水解酶
hydrolysis	水解作用
ileus	肠梗阻
intermediary product	中间产物
ionic strength	离子强度
Isoamylase	异淀粉酶
lactate dehydrogenase	乳酸脱氢酶
ligand	配位体
ligase	连接酶类
lipases	酯酶
liquefacient	液化剂
lyase	裂合酶类
macroamylasemia	巨淀粉酶血症
malolactic	苹果乳酸的
maltase	麦芽糖酶
maltodextrin	麦芽糖糊精
mannosidases enzymes	甘露糖苷酶
mayonnaise	蛋黄酱
metabolism	新陈代谢
mitochondrial	线粒体的
mump	腮腺炎
myosin	肌球蛋白
nomenclature	命名法
oligogalacturonic acid	臭氧低半乳糖醛酸
oligosaccharides	低聚糖
orthophenanthroline	邻二氮菲
ovarian cyst	卵巢囊肿

oxidoreductase	氧化还原酶
parameter	参数
pectate lyase	果胶裂解酶
pectin methylesterase	果胶甲酯酶
pepstatin	胃蛋白酶抑制剂
phaseolamin	菜豆素
polygalacturonase	聚半乳糖醛酸酶
prosthetic	辅基的
proteases	蛋白酶
protomer	原体
prototropic group	质子群
pullulanase	支链淀粉酶
pyrophosphate	焦磷酸盐
reaction specificity	反应特性
renin	肾素
retrotransposon	反转录专座子
retroviral	逆转录病毒的
saccharification	糖化作用
sarcoplasm	肌质
secondary and tertiary conformation	二级和三级结构
serine	丝氨酸
serum amylase	血清淀粉酶
strangulation	窒息
substrate specificity	底物特性
subtilisin	枯草杆菌蛋白酶
thioester	硫酯
transferase	转移酶
triglyceride	甘油三酯
triosephosphate	磷酸丙糖
triphosphate	三磷酸盐

FOOD CHEMISTRY

Chapter 8 Colorants

8.1 Introduction

To understand colorants in foods some terms need to be defined. Color refers to human perception of colored materials—red, green, blue, etc. A colorant is any chemical, either natural or synthetic, which imparts color. Foods have color because of their ability to reflect or emit different quantities of energy at wavelengths able to stimulate the retina in the eye. The energy range to which the eye is sensitive is referred to as visible light. Visible light, depending on an individual's sensitivity, encompasses wavelengths of approximately 380-770 nm. This range makes up a very small portion of the electromagnetic spectrum (Fig. 8.1). In addition to obvious colors (hues), black, white, and intermediate grays are also regarded as colors.

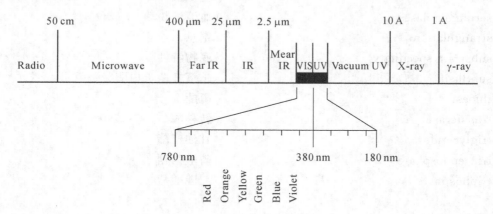

Fig. 8.1 Electromagnetic spectrum

Pigments are natural substances in cells and tissues of plants and animals that impart color. Dyes are any substances that lend colors to materials. The term "dye" is commonly used in the textile industrial. In the food industry, a dye is a food-grade water-soluble colorant certified by the government. Only those specific dyes that are certified as food colors may be used in foods, drugs, and cosmetics. Added to the approved list of certified colors are the lakes. Lakes are dyes extended on a substratum and they are oil dispersible.

The dye/substratum combination is achieved by adsorption, co-precipitation, or chemical reaction. The complex involves a salt of a water-soluble primary dye and an approved insoluble base stratum. Alumina is the only approved substratum for preparing food applicable lakes in most countries. Colorants exempt from certification may also be used. These are natural pigments, for example, anthocyanin, juice concentrate, annatto extract, or substances synthesized, but identical to the natural pigment, for example β-carotene.

Color and appearance are the major, if not the most important, quality attributes of foods. It is because of its easy being perceived by consumer when purchasing foods. One can provide consumers the most nutrition, safest, and most economical foods, but if they are not color attractive, purchase will not occur. The consumer also relates specific colors of foods to quality. Specific colors of fruits are often associated with maturity, redness of raw meat is associated with freshness, a green apple may be judge immature (although some are green when ripe), and brownish-red meat as not fresh.

Color also influences flavor perception. The consumer expects red drinks to be strawberry or cherry flavored, yellow to be lemon, and green to be lime flavored. It should also be noted that some substances such as β-carotene or riboflavin are not only colorants but nutrients as well. It is clear therefore that the color of foods has multiple effects on consumers, and it is wrong to regard color as being purely cosmetic.

Many food pigments are, unfortunately, unstable during processings and storages. Prevention of the undesirable changes of pigments is usually difficult or impossible. Depending on the specific pigment, stability is impacted by the factors such as the presence or absence of light, oxygen, heavy metals, and oxidizing or reducing agents; temperature, water activity; and pH. Because of the instability of pigments, colorants are sometimes added to foods.

The purpose of this chapter is to provide an understanding of colorant chemistry, concentrated on the natural pigments from plants and animal tissues, that is the essential prerequisite for controlling the color and color stability of foods.

8.2 Natural Pigments

8.2.1 Chlorophyll

8.2.1.1 Structure of Chlorophyll

Chlorophylls are the major pigments in green plants, algae, and photosynthetic bacteria. They are magnesium complexes derived from the porphin. The porphin, also named "tetrapyrrole", is a fully unsaturated macrocyclic structure that contains four pyrrole rings linked by single bridging carbons. Substituted porphins are named porphyrins.

Several chlorophylls are found in nature. Their structures differ in the substituents around the porphin nucleus. Chlorophyll a and b are found in green plants in an approximate ratio of 3 : 1. They differ in the carbon C-3 substituent. Chlorophyll a contains a methyl group while chlorophyll b contains a formyl groups (Fig. 8.2). Both chlorophylls have a vinyl and an ethyl group at the C-2 and C-4 position, respectively; a carbomethoxy group at the C-10 position of the isocylic rings, and a phytol group esterfied to propionate at the C-7 position. Phytol is a 20-carbon monounsaturated isoprenoid alcohol.

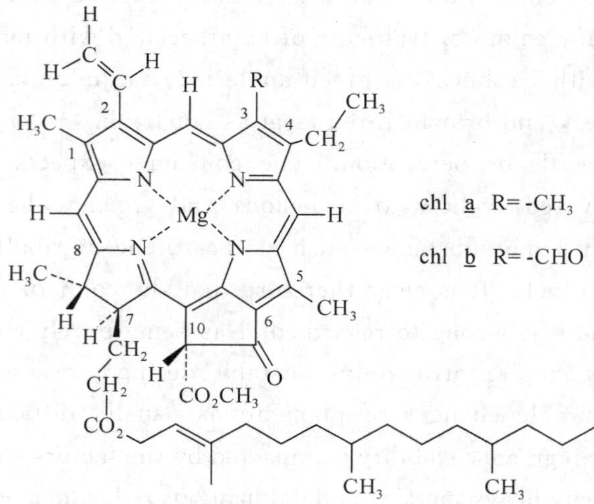

Fig. 8.2 Structure of chlorophyll

Chlorophylls are located in the lamellae of intercellular organelles of green plants known as chloroplasts. They are associated with carotenoids, lipids, and lipoproteins. The bonds between these molecules are weak (non-covalent bonds) and easily broken; hence chlorophylls can be completely extracted by organic solvents such as acetone, methanol, ethanol, ethyl acetate, pyridine, and dimethylformamide. Nonpolar solvents such as hexane or petroleum ether are less effective.

8.1.1.2 Alterations of Chlorophyll

Enzymatic

Chlorophyllase is the only enzyme known to catalyze the degradation of chlorophyll. Chlorophyllase is an esterase, it catalyzes cleavage of phytol from chlorophylls and its Mg-free derivatives (pheophytins) *in vitro*, forming chlorophyllides and pheophorbides, respectively (Fig. 8.3). The enzyme is active in solutions containing water, alcohols, or acetone. Formation of chlorophyllides in fresh leaves does not occur until the enzyme has been heat activated postharvest. The optimum temperature for chlorophyllase activity in vegetables runs between 60 ℃ and 80 ℃. Enzyme activity decreases when plant tissue is heated above 80 ℃, and chlorophyllase loses its activity if heated to 100 ℃.

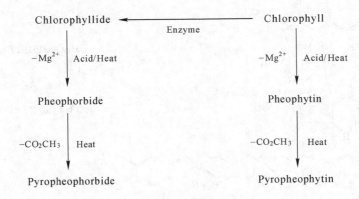

Fig. 8.3 Chlorophyll and its derivatives

Heat and Acid

Chlorophyll derivatives formed during heating or thermal processing can be classified into two groups based on the presence or absence of the magnesium atom in the tetrapyrrole center. Mg-containing derivatives are green in color, while Mg-free derivatives are olive-brown in color. The latter are chelators and when, for example, sufficient zinc or copper ions are available they will form green zinc or copper complexes.

The chlorophyll degradation in heated vegetable tissues is affected by tissue pH. In a basic media (pH 9.0) chlorophyll is very stable, toward heat, whereas in an acidic media (pH 3.0) it is unstable. A decrease of 1 pH unit can occur during heating of plant tissue through the release of acids, and this has an important detrimental effect on the rate of the chlorophyll degradation. It was proposed that the pheophytin formation in plant cells is initiated by a heat-induced increase in permeability of hydrogen ions across cell membranes. The critical temperature for the initiation of the pheophytin formation coincided with gross changes in the membrane organization.

The replacement of the C-10 carbomethoxy group of pheophytin with a hydrogen atom results in the formation of olive-colored pyropheophytin. The wavelengths of maximum light absorption by pyropheophytin are identical to those for pheophytin. The chlorophyll alteration during heating is sequential, and proceeds according to the kinetic sequence shown in Fig. 8.4.

8.2.1.3 Technology of Color Preservation

Efforts to preserve green color in canned vegetables have concentrated on retaining chlorophyll, forming or retaining green derivatives of chlorophyll, that is, chlorophyllides, or creating a more stable green color through the formation of metal complexes.

Acid Neutralization to Retain Chlorophyll

The addition of alkalizing agents to canned green vegetables can result in improved

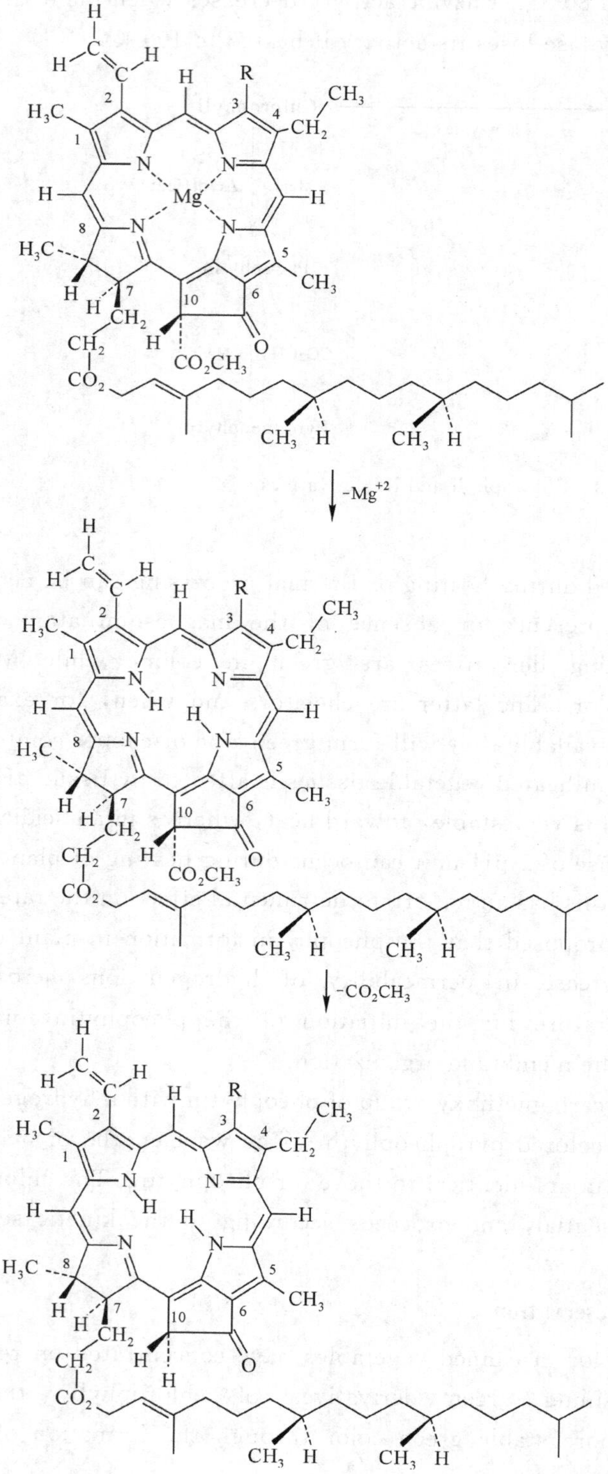

FOOD CHEMISTRY

CHLOROPHYLL (chl)

chl *a.* R= –CH$_3$

chl *b.* R= –CHO

PHEOPHYTIN (phe)

phe *a.* R= –CH$_3$

phe *b.* R= –CHO

PYROPHEOPHYTIN (pyro)

pyro *a.* R= –CH$_3$

pyro *b.* R= –CHO

Fig. 8.4 Formation of pheophytin and pyropheophytin from chlorophyll

retention of chlorophylls during processings. Techniques have involved the addition of calcium oxide and sodium dihydrogen phosphate in blanch water to maintain product pH or to raise the pH to 7.0. Magnesium carbonate or sodium carbonate in combination with sodium phosphate has been tested for this purpose. However, all of these treatments result in softening of the tissue and an alkaline flavor.

Blair in 1940 recognized the toughening effect of calcium and magnesium when added to vegetables. This observation lead to the use of calcium or magnesium hydroxide for the purpose of raising pH and maintaining texture. This combination of treatments became known as the "Blair process". The commercial application of these processes has not been successful because of the inability of the alkalizing agents to effectively neutralize interior tissue acids over a long period of time, resulting in substantial color loss after less than 2 months of storages, and increasing the pH of canned vegetables can also cause hydrolysis of amides such as glutamine or asparagine with formation of undesirable ammonia-like odors. In addition, fatty acids formed by the lipid hydrolysis during high pH blanching may oxidize to form rancid flavors.

High-Temperature Short-Time processing

Commercially sterilized foods processed at a higher than normal temperature for a relatively short time (HTST) often exhibit better retention of vitamins, flavor and color than do conventionally. The greater retention of these constituents in HTST foods results because their destruction is more temperature dependent than that for the inactivation of Clostridium botulinum spores.

Other studies to preserve the green color of processed vegetables was to combine HTST processing with pH adjustment. Samples treated in this manner were initially greener and contained more chlorophyll than control samples (typical processing and pH). However, the improvement in color was generally lost during storage.

Enzymatic Conversion of Chlorophyll to Chlorophyllides to Retain Green Color

Blanching at lower temperatures than conventionally used to inactivate enzymes has been suggested as a means of achieving better retention of color in green vegetables. Early studies showed that the better color of processed spinach blanched under low temperature conditions (65 ℃ for up to 45 min) was caused by the heat-induced conversion of chlorophyll to chlorophyllides by the enzyme chlorophyllase. However, the improvement in color retentions achieved by this approach was insufficient to warrant commercialization of the process.

Commercial Application of Metal Complex

Current efforts to improve the color of green processed vegetables and to prepare chlorophylls that might be used as food colorants/lave involved the use of either zinc or copper complexes of chlorophyll derivatives. Copper complexes of pheophytin and pheophorbide are available commercially under the names copper chlorophyll and copper chlorophyllin, respectively. The chlorophyll derivatives cannot be used in foods in the

United States. Their use in canned foods, soups, candy, and dairy products is permitted in most European countries under regulatory control of the European Economic Community. The Food and Agriculture Organization (FAO) of the United Nations has certified their use as safe in foods, provided no more than 200 mg/kg of free ionizable copper is present. The copper complexes have greater stability than comparable Mg complexes; for example, after 25 hr at 25 ℃, 97% of the chlorophyll degrades while only 44% of the copper chlorophyll degrades.

Regreening of Thermal Processed Vegetables

It has been observed that when vegetables purees are commercially sterilized, small bright-green areas occasionally appear. It was determined that pigments in the bright green areas contained zinc and copper. This formation of bright-green areas in vegetable purees was termed "regreening". Regreening of commercially processed vegetables has been observed when zinc and/or copper ions are present in process solutions. Okra when processed in brine solution containing zinc chloride retains its bright green color, and this is attributed to the formation of zinc complexes of chlorophyll derivatives.

8.2.2 Heme Compounds

Heme pigments are responsible for the color of meat. Myoglobin is the primary pigment of red meat (muscle tissue) and hemoglobin is the pigment of blood. Most of the hemoglobin is removed when animals are slaughtered and bled. Thus, myoglobin is responsible for 90% or more of the pigmentation in properly bled red meat. The myoglobin quantity varies considerably among muscle tissues and is influenced by species, age, sex, and physical activity of the animals. For example, pale-colored veal has lower myoglobin content than red-colored beef. Muscle-to-muscle differences within an animal are also apparent, and these differences are caused by varying quantities of myoglobin presented within the muscle fibers. Such is the case with poultry, where light-colored breast muscle is easily distinguished from the dark muscle color of leg and thigh muscles. The myoglobin content of selected muscles are shown in Table 8.1.

Table 8.1 The myoglobin content of selected muscles

Myoglobin	mg/g
Chicken white meat	0.05
Chicken dark meat	1-3
Pork and veal	1-3
PSE pork	1-3
Beef (A maturity)	4-10
Cow beef	15-20
Beef hearts	20-30
Whale muscle	40

The major pigments found in fresh, cured, and cooked meat are listed in Table 8.2. Other minor pigments present in muscle tissue include the cytochrome enzymes, flavins, and vitamin B_{12}.

8.2.2.1 Structures of Heme Compounds

Myoglobin is a globular protein consisting of a single polypeptide chain. The protein portion of the molecule known as globin is comprised of 153 amino acids. The chromophore component is a porphyrin known as heme. Within the porphyrin ring, a centrally located iron atom is complexed with tetra-pyrrole nitrogen atoms. Thus, myoglobin is a complex of globin and heme. The heme porphyrin is present within a hydrophobic pocket of the globin and bound to a histidine residue (Fig. 8.4). The centrally located iron atom possesses six coordination sites, four of which are occupied by the nitrogen atoms within the tetra-pyrrole ring. The fifth coordination site is bound by the histidine residue of globin, leaving the sixth site available to complex with electronegative atoms donated by various ligands, for example oxygen.

Table 8.2 Major pigments found in fresh, cured, and cooked meat

Pigment	Mode of formation	State of iron	State of hematin nucleus	State of globin	Color
1. Myoglobin	Reduction of metmyoglobin; deoxygenation of oxymyoglobim	Fe^{2+}	Intact	Native	Purplish-red
2. Oxymyoglobim	Oxygenation of myoglobin	Fe^{2+}	Intact	Native	Bright red
3. Metmyoglobim	Oxidation of myoglobin, oxymyoglobim	Fe^{3+}	Intact	Native	Brown
4. Nitric oxide myoglobim (nitrosometmyoglobim)	Combination of metmyoglobim with nitric oxide	Fe^{3+}	Intact	Native	Crimson
5. Nitric oxide metmyoglobim (nitrosometmyoglobim)	Combination of metmyoglobim with nitric oxide	Fe^{3+}	Intact	Native	Crimson
6. Metmyoglobim nitrite	Combination of metmyoglobim with excess nitrite	Fe^{3+}	Intact	Native	Reddish-brown
7. Globin myohemochromogen	Effect of heat, denaturing agents on myoglobim, oxymoglobim; irradiation of globin hemichrogmogen	Fe^{2+}	Intact (usually bound to denatured protein other than globin)	Denatured (usually detached)	Dull red
8. Globin myohemichromogen	Effect of heat, danaturing agents on myoglobim, oxymoglobim, metmyoglobim, hemochromogen	Fe^{3+}	Intact (usually bound to denatured protein other than globin)	Denatured (usually detached)	Brown (sometimes greyish)
9. Nitric oxide myohemochromogen	Effect of heat, denaturing agents on nitric oxide myoglobin	Fe^{2+}	Intact	Denatured	Bright red (pink)
10. Sulfmyoglobin	Effect of H_2S and oxygen on myoglobin	Fe^{3+}	Intact but one double bond saturated	Native	Green
11. Metsulfmyoglobin	Oxidation of sulfmyoglobin	Fe^{3+}	Intact but one double bond saturated	Native	Red
12. Choleglobin	Effect of hydrogen peroxide on myoglobin or oxymyoglobin; effect of ascorbine or other reducing agent on oxymyoglobin	Fe^{2+} or Fe^{3+}	Intact but one double bond saturated	Native	Green
13. Nitrihemin	Effect of large excess nitrite and heat on 5	Fe^{3+}	Intact but reduced	Absent	Green
14. Verdohaem	Effect of reagents as in 7-9 in excess	Fe^{3+}	Porphyrin ring opened	Absent	Green
15. Bile pigments	Effect of reagents as in 7-9 in large excess	Fe absent	Porphyrin ring destroyed Chain of porphyrins	Absent	Yellow or colorless

FOOD CHEMISTRY

Hemoglobin consists of four myoglobins linked together as a tetramer. It is a component of red blood cells, forming reversible complexes with oxygen in the lung.

This complex is distributed via the blood to various tissues throughout the animal where oxygen is absorbed. It is the heme group that binds molecular oxygen. Myoglobin within the cellular tissue acts in a similar mode with hemoglobin when it carries oxygen. Thus, myoglobin makes oxygen stored within the tissues, and available for metabolism.

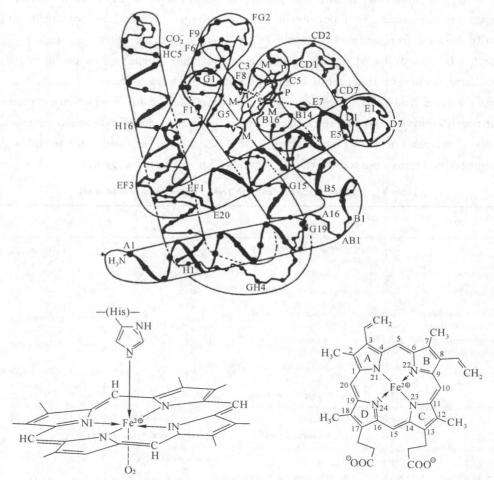

Fig. 8.4 Primary(left) and tertiary(right) structure of myoglobin

8.2.2.2 Chemistry and Alterations of Color

Different reactions as shown in Fig. 8.5 can cause color alterations of myoglobin in fresh and cured meat as listed in Table 8.2. Hydrogen peroxide can react with either the ferrous or ferric site of heme, resulting in green-colored choleglobin. Also, in the presence of hydrogen sulfide and oxygen, green sulfomyoglobin can form. It is thought that hydrogen peroxide and/or hydrogen sulfide come from bacterial growth.

Reactions occur in cured meats are responsible for the stable pink color of cured meat products. The first reaction occurs between nitric oxide (NO) and myoglobin to produce nitric oxide myoglobin (NOMb). also know as nitrosylmyoglobin. NOMb is bright red and unstable. Upon heating, the more stable nitric oxide myohemochromogen (nitrosylhemochrome) forms. This product yields the desirable pink color of cured meats. If metmyogthbin is present, it has been postulated that reducing agents are required to convert metmyoglobin to myoglobin before the reaction with nitric oxide can take place. Alternatively, nitrite can interact directly with metmyoglobin, in the presence of excess nitrous acid, nitrimetmyoglobin (NMMb) will form. Upon heating in a reducing environment, NMMb is converted to nitrihemin, a green pigment.

In the absence of oxygen, nitric oxide complexes of myoglobin are relatively stable. However, under aerobic conditions, these pigments are sensitive to light. If reductants such as ascorbate or sulftlydryl compounds are added, the reductive conversion of nitrite to nitric oxide is favored. Thus, under these conditions, formation of nitric oxide myoglobin occurs more readily.

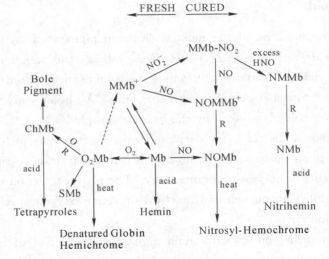

Fig. 8.5 Myoglobin reactions in fresh and cured meats

ChMb=cholemyglobin (oxidized porphyrin ring);
O_2Mb=oxymyoglobin (Fe^{2+});
MMb=metmyoglobin (Fe^{3+});
Mb=myoglobin (Fe^{2+});
$MMb-NO_2$=metmyoglobin nitrite;
NOMMb=nitrosylmetmyoglobin;
NOMb=nitrosylmyoglobin;
NMMb=nitrometmyoglobin;
NMb=nitromyoglobin, the latter two being reaction products of nitrous acid and the heme portion of the molecule;
R=reductant;
O=strong oxidizing conditions.

8.2.2.3 Stability of Meat Pigments

Many factors operative in food system can influence the stability of meat pigments. Some environmental conditions that have important effects on meat color and pigment stability include exposure to light, temperature, relative humidity, pH, and the presence of specific bacterias.

Some specific reactions, such as lipid oxidations, are known to increase the rate of pigment oxidations. Similarly, the color stability can be improved by the addition of antioxidants such as ascorbic acid, vitamin E, butytated hydroxyanisole (BHA), propyl gallate or natural compounds such as tea or bamboo extracts. These compounds have been shown to delay the lipid oxidation and improve retention of color in meat tissues. Other biochemical factors, such as the rate of oxygen consumption prior to slaughter and activity of metmyoglobin reductase, can influence the color stability of fresh meat.

8.2.3 Carotenoids

Carotenoids are one kind of the most widespread pigments in nature. Edible plant tissues contain a wide variety of carotenoids. Red, yellow, and orange fruits, root crops, and vegetables are rich in carotenoids. Prominent examples include tomatoes, carrots, red peppers, pumpkins, squashes, corn and sweet potatoes. All green leafy vegetables contain carotenoids but their color is masked by the green chlorophylls.

It has been known that carotenoids play important functions in photosynthesis and photoprotection in plant tissues. In all chlorophyll-containing tissues, carotenoids function as secondary pigments in harvesting light energy. The photoprotection role of carotenoids stems from their ability to quench and inactivate reactive oxygen species formed by the exposure to light and air.

The most prominent role of carotenoid pigments in the diet of humans and other animals is their ability to serve as precursors of vitamin A. Among the carotenoids, beta-carotene possesses the greatest provitamin A activity because of its two β-ionone rings (Fig. 8.6), other commonly seen carotenoids, such as α-carotene and β-cryptoxanthin, also possess provitamin A activity. In 1981, Scientists found that consumption of fruits and vegetables high in carotenoid content is associated with a decreased incidence of specific cancers in humans. Recently, interest has focused on the presence of processing-induced cis isomeric carotenoids in the diet and on their physiological significances.

8.2.3.1 Structure of Carotenoids

Carotenoids are comprised of two structural groups: the hydrocarbon carotenes and the oxygenated xanthophylls. Oxygenated carotenoids (xanthophylls) consist of a variety of derivatives frequently containing hydroxyl, epoxy, aldehyde, and keto groups. In addition, fatty acid esters of hydroxylated carotenoids are also widely found in nature. Thus, more than 5600 carotenoids have been identified and compiled. Furthermore, when

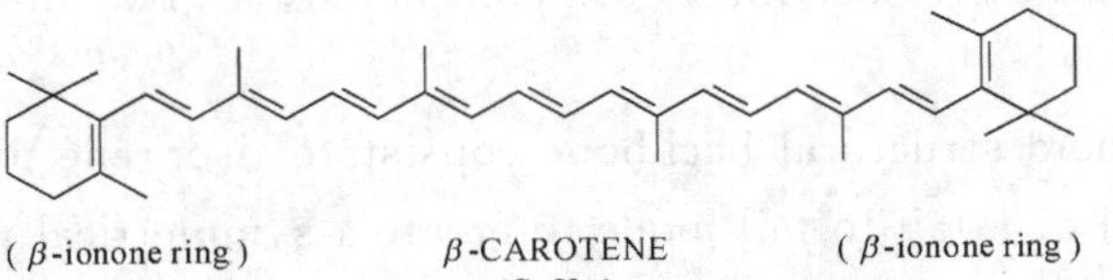
(β-ionone ring) β-CAROTENE (β-ionone ring)
($C_{40}H_{56}$)

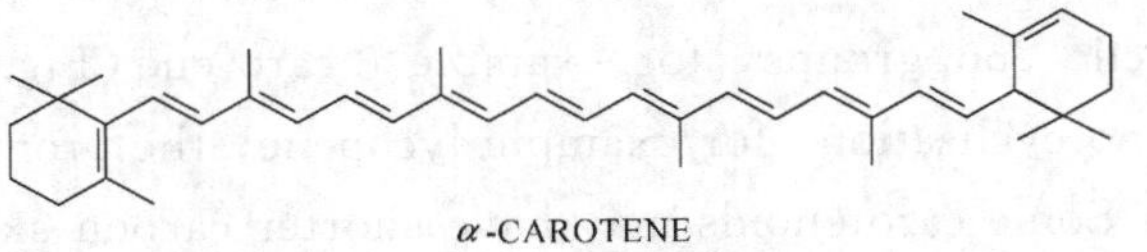
α-CAROTENE
($C_{40}H_{56}$)

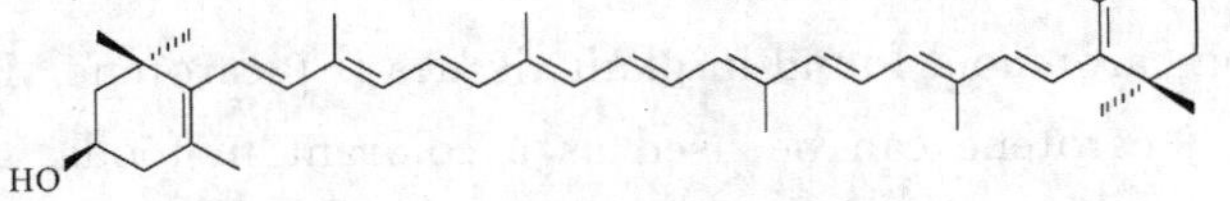
β-CRYPTOXANTHIN
($C_{40}H_{56}O$)

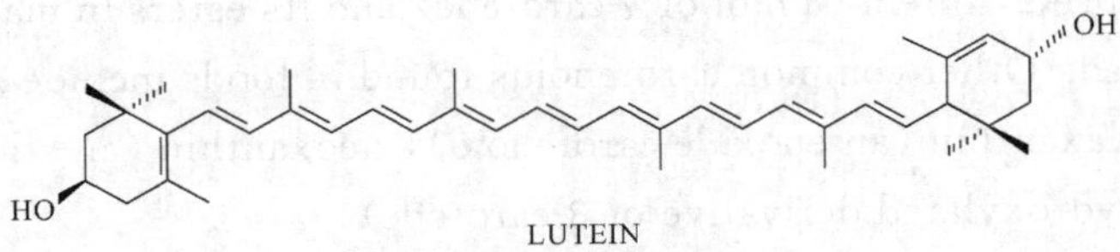
LUTEIN
($C_{40}H_{56}O_2$)

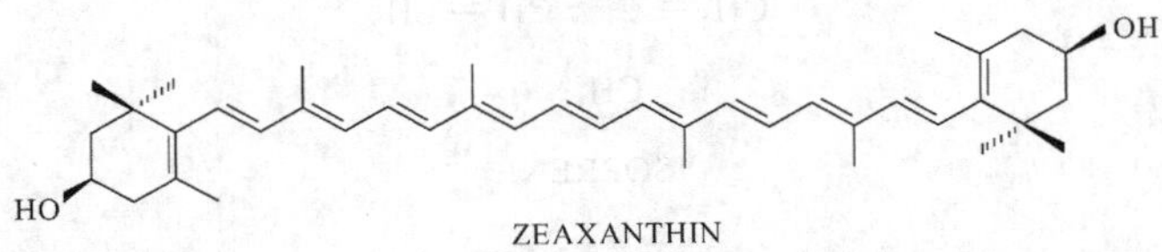
ZEAXANTHIN
($C_{40}H_{56}O_2$)

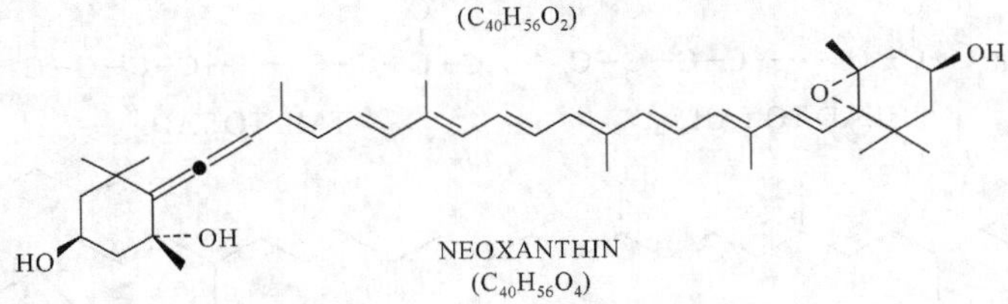
NEOXANTHIN
($C_{40}H_{56}O_4$)

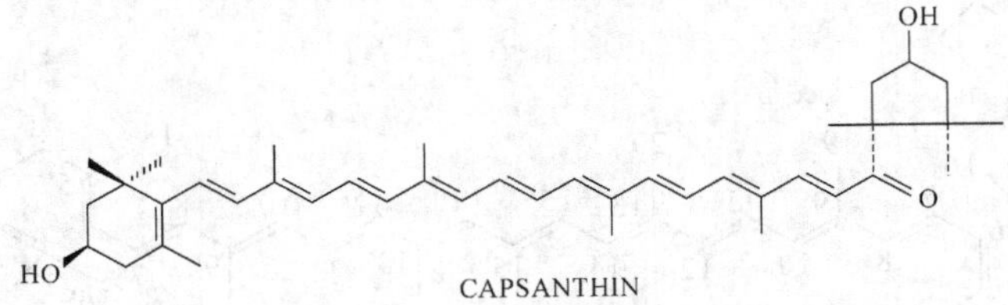
CAPSANTHIN
($C_{40}H_{56}O_3$)

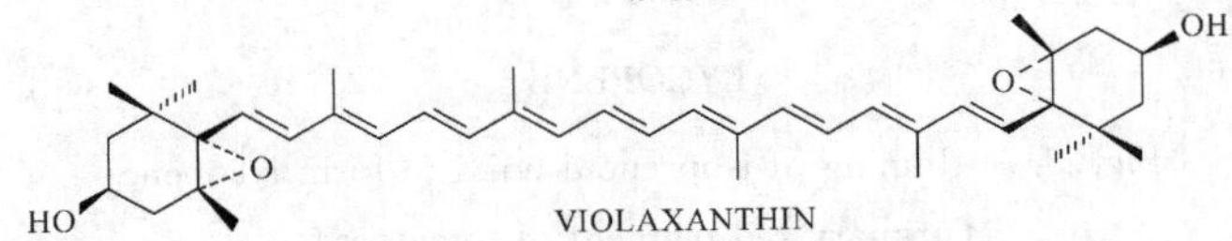
VIOLAXANTHIN
($C_{40}H_{56}O_4$)

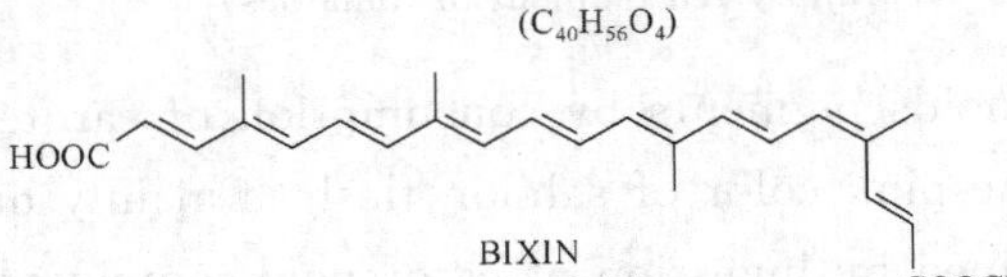
BIXIN
($C_{25}H_{30}O_4$)

Fig. 8.6 Structures of commonly occurring carotenoids

FOOD CHEMISTRY

geometric isomers of *cis or tranns* forms are considered, a great many configurations are possible.

The basic carotenoid structural backbone consists of isoprene units linked covalently either in a head-to-tail or a tail-to-tail mode to create a symmetrical molecule (Fig. 8.7). Other carotenoids are derived from this primary structure of forty carbons. Some structures contain cyclic end groups, for example β-carotene (Fig. 8.6) while others possess either one or no cyclization, for example lycopene, the prominent red pigment in tomatoes (Fig. 8.7). Some carotenoids may have shorter carbon skeletons, for example bixin, a yellow pigment from annatto seeds.

The most common carotenoid found in plant tissues is β-carotene. Both the naturally derived and synthetic β-carotene can be used as a colorant in foods. Some carotenoids found in plants are shown in Fig. 8.6, they include α-carotene in carrots, capsanthin in red peppers and paprika, lutein (a diol of α-carotene) and its esters in marigold petals, and bixin in annatto seed. Other common carotenoids found in foods include zeaxanthin (a diol of β-carotene), violaxanthin (an epoxide carotenoid), neoxanthin (an allenic triol), and β-cryptoxanthin (a hydroxylated derivative of β-carotene).

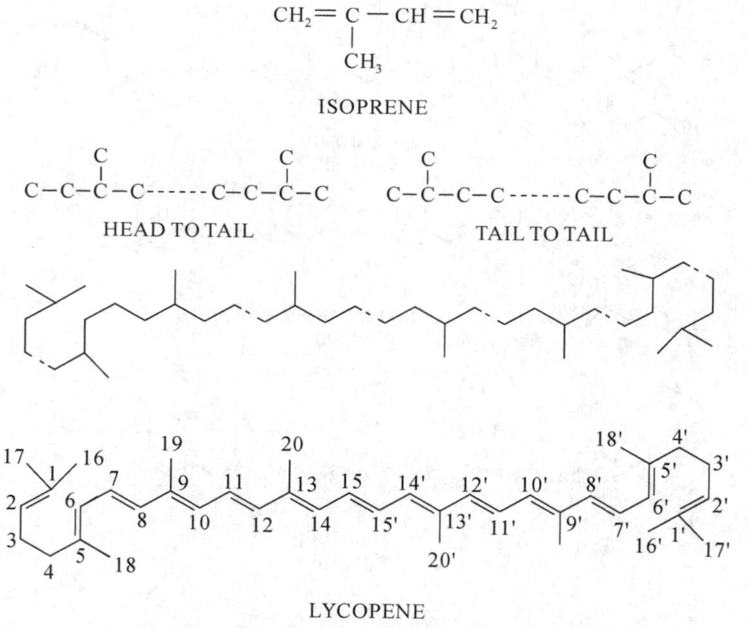

$$CH_2=C-CH=CH_2$$
$$|$$
$$CH_3$$

ISOPRENE

Fig. 8.7 Joining of isoprenoid units to form lycopene
(primary red pigment of tomatoes)

Animals derive carotenoids pigments by consumption of carotenoid-containing plant materials. For example, the pink color of salmon flesh is mainly due to the presence of astaxanthin, which is obtained by ingestion of carotenoid-containing marine plants. It is also well known that some carotenoids in both plants and animals are bound to or associated with proteins. The red astaxanthin pigment of shrimp and lobster exoskeletons is blue in color when complexed with proteins. Heating denatures the protein in the complex and alters the visual properties of the pigment, thus the color change from blue to red. Other unique structures include carotenoid glycosides, some of which are found in bacterias and other microorganisms.

8.2.3.2 Physical and Chemical Propeties

All classes of carotenoids including hydrocarbons and oxygenated xanthophylls are lipophilic and are soluble in oils and organic solvents. Extraction procedures of carotenoids from tissue utilize organic solvents that must penetrate a hydrophilic matrix. Hexane-acetone mixtures are commonly employed for this purpose, but special solvents and treatments are sometimes needed to achieve satisfactory separations. Since they range in color from yellow to red, detection wavelengths for monitoring carotenoids typically range from approximately 430 to 480 nm. Higher wavelengths are usually used for some xanthophylls to prevent interferences from chlorophylls.

Carotenoids are moderately heat stable and are easily oxidized to lose color by oxidation because of their highly conjugated unsaturated structures. The stability of a particular carotenoid to oxidation is highly dependent on its environment. Within tissues, the carotenoids are often compartmentalized and protected from oxidations. However, physical damage to the tissue or extraction of the carotenoids increases their susceptibility to oxidations. Furthermore, storage of carotenoid pigments in organic solvents will often accelerate decompositions. The products of their degradations are very complex and largely uncharacterized except for β-carotene. During the oxidation of β-carotene, epoxides and carbonyl compounds are initially formed. Further oxidation results in formations of short-chain mono-and dioxygenated compounds including epoxy-β-ionone. For provitamin A carotenoids, epoxide formation in the ring results in loss of the provitamin activity. Extensive oxidations will result in the loss of color. Oxidative destructions of β-carotene are intensified in the presence of sulfite and metal ions. Enzymatic activity, particularly lipoxygenase, hastens oxidative degradation of carotenoids.

Carotenoids can be easily isomerized by heat, acid, or light. In general, the conjugated double bonds of carotenoid compounds exist in an all-trans configuration. The cis isomers of a few carotenoids can be found naturally in plant tissues, especially in algae, which are currently being harvested as a source of carotenoid pigments. Isomerization reactions are readily induced by thermal treatments, exposure to organic solvents and light, treatment with acids, and particularly with iodine. Theoretically, large numbers of possible geometrical configurations could result from isomerization because of the extensive number of double bonds present in carotenoids. Cis/trans isomerization affects the provitamin A activity of carotenoids. The provitamin A activity of β-carotene cis isomers ranges, depending on the isomeric form, from 13% to 53% as compared to that of all-trans-β-carotene. The reason for this reduction in provitamin A activity is not known.

8.2.3.3 Antioxidant Activity

Carotenoids have antioxidant properties because they can be readily oxidized. In addition to cellular and *in vitro* protection against singlet oxygen, carotenoids, at low oxygen partial pressures, inhibit lipid peroxidations. But at high oxygen partial pressures, β-carotene has pro-oxidant properties. In the presence of molecular oxygen, photosensitizers (i.e., chlorophyll), and light, singlet oxygen may be produced, which is a highly reactive oxygen species. Carotenoids are known to quench singlet oxygen and thereby protect against cellular oxidative damages. Not all carotenoids are equally effective as photochemical protectors. Lycopene, for example, is known to be especially efficient in quenching singlet oxygen relative to other carotenoid pigments. When cis isomers are created, only slight spectral shifts occur and thus color of the product is mostly unaffected; however, a decrease in provitamin A activity occurs.

8.2.3.4 Stability during Processing

Carotenoids are relatively stable during typical processing and storage conditions of most fruits and vegetables. However, blanching is known to influence the level of carotenoids. Often blanched plant products exhibit an apparent increase in carotenoid content relative to raw tissues. This is caused by inactivation of lipoxygenase, which is known to catalyze oxidative decomposition of carotenoids. Lye peeling, which is commonly used for orange and sweet potatoes, causes little destruction or isomerization of carotenoids.

Although carotene historically has been regarded as fairly stable during heating, it is now known that the heat sterilization induces cis/trans isomerization reactions. To lessen excessive isomerization, the severity of thermal treatments should be minimized when possible. In the case of high-temperature heating in oils, not only will carotenoids isomerize but thermal degradation will also occur. Products arising from severe heating of β-carotene in the presence of air are similar to those arising from severe heating of β-carotene oxidation. In contrast, extensive degradation of carotenoids will occurs when exposed to oxygen especially *in vitro*.

8.2.4 Phenolic compound

Phenolic compounds comprise a large group of substances, and flavonoids are an important subgroup. The flavonoid subgroup contains the anthocyanins, one of the most broadly distributed pigment groups in the plant world. Anthocyanins are responsible for a wide range of colors in plants, including blue, purple, violet, magenta, red, and orange.

8.2.4.1 Anthocyanins

Anthocyanins are considered flavonoids because of the characteristic $C_6 C_3 C_6$ carbon skeleton. The basic chemical structure of the flavonoid group and the relationship to antbocyanin are shown in Fig. 8.8. Within each group there are many different compounds with their color depending on the substituents on rings A and B.

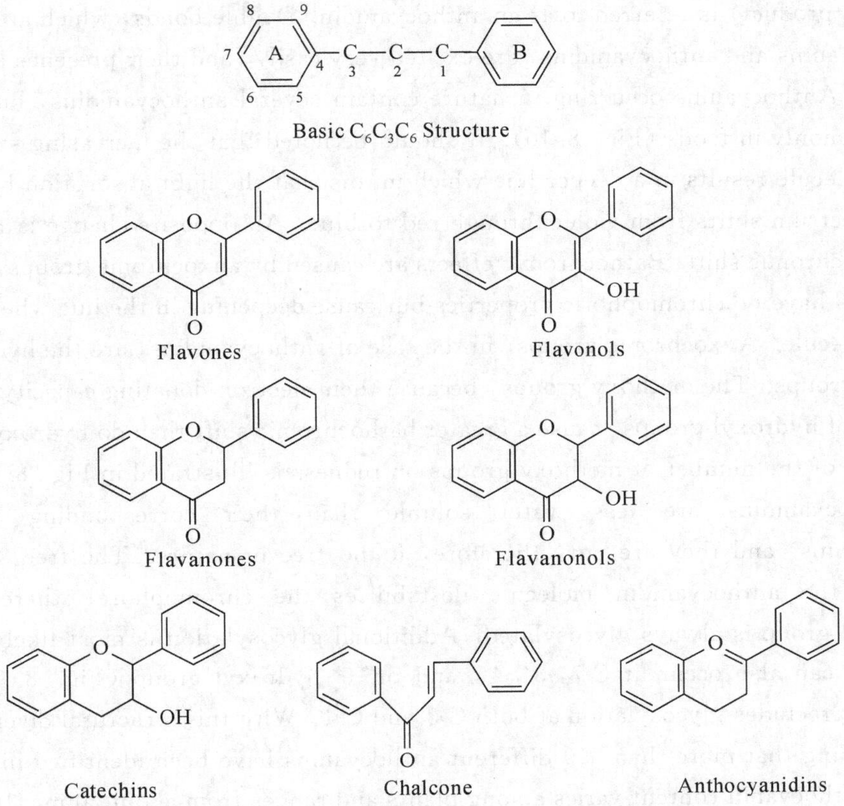

Fig. 8.8 Carbon skeleton of the flavonoid group

Structure

The base structure of anthocyanin is the 2-phenylbenzopyrylium of flavylium salt (Fig. 8.9). Anthocyanins exist as glycosides of polyhydroxy and/or polymethoxy derivatives of the salt. Anthocyanins differ in the number of hydroxyl and/or methoxy groups present, the types, numbers, and sites of attachment of sugars to the molecule. The most common sugars are glucose, galactose, arabinose, xylose, and homogenous or heterogeneous di-and trisaccharides formed by combining these sugars.

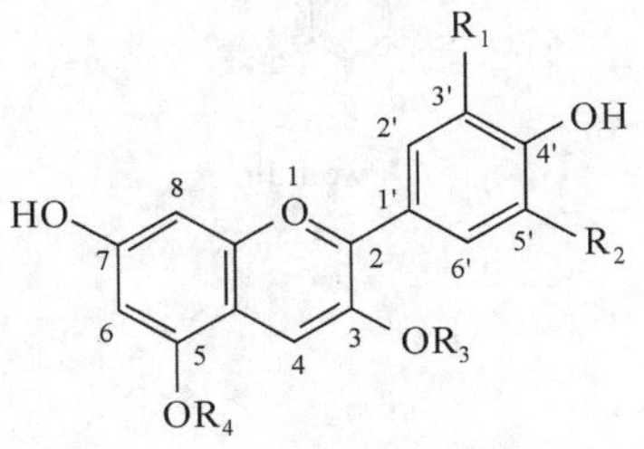

Fig. 8.9 The flavylium cation
R_1 and $R_2 =$ -H, -OH, or -OCH$_3$, $R_3 =$ -glycosyl, $R_4 =$ -H or -glycosyl.

When the sugar moiety of an anthocyanin is hydrolyzed, the aglycone (the nonsugar hydrolysis product) is referred to as an anthocyanidin. Double bonds, which are abundant in anthocyanins and anthocyanidins, are excited very easily, and their presence is essential for color. Anthocyanins occurring in nature contain several anthocyanidins, but only six occur commonly in foods (Fig. 8.10). It should be noted that the increasing substitution on the molecule results in a deeper hue which means that the light absorption band in the visible spectrum shifts from violet through red to blue. An opposite change is referred to as a hypsochromic shift. Bathochromic effects are caused by auxochrome groups, which by themselves have no chromophoric properties but cause deepening in the hue when attached to the molecule. Auxochrome groups, in the case of anthocyanidins, are the hydroxyl and methoxy groups. The methoxy groups, because their electron-donating capacity is greater than that of hydroxyl groups, cause a greater bathochromic shift than do hydroxyl groups. The effect of the number of methoxy groups on redness is illustrated in Fig. 8.10.

Anthocyanidins are less water soluble than their corresponding glycosides (anthocyanins) and they are not, therefore, found free in nature. The free 3-hydroxyl group in the anthocyanidin molecule destabilizes the chromophore; therefore, the 3-hydroxyl group is always glycosylated. Additional glycosylation is most likely to occur at C-5 and can also occur at C-7,-3',-4', and/or -5' hydroxyl group (Fig. 8.9). Steric hindrance precludes glycosylation at both C-3' and C-4'. With this structural diversity, it is not surprising that more than 250 different anthocyanins have been identified in the plant world. Anthocyanin content varies among plants and ranges from about 20 mg/100 g fresh weight to as high as 600.

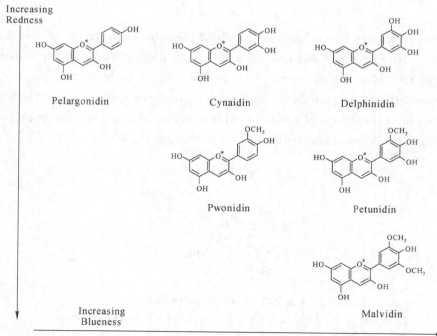

Fig. 8.10   Most common anthocyanidins in foods, arranged in increasing redness and blueness

Color and Stability of Anthocyanins

Anthocyanin pigments are relatively unstable, with greatest stability occurring under acidic conditions. Both the hue of the pigment and its stability are greatly impacted by substituents on the aglycone. Degradation of anthocyanins easily occurs during extractions from plant tissues and during processings and storages of foods tissues. Major factors governing degradation of anthocyanins are pH, temperature, and oxygen concentrations. Factors that are usually of less importance are the presence of degradative enzymes, ascorbic acid, sulfur dioxide, metal ions, and sugars.

Structural Transformation and pH

Degradation rates vary greatly among anthocyanins because of their diverse structures. Generally, increased hydroxylation decreases stability, while increased methylation increases stability. The color of foods containing anthocyanins that are rich in pelargonidin, cyanidin, or delphinidin aglycones is less stable than that of foods containing anthocyanins that are rich in petunidin or malvidin aglycones. The increased stability of the latter group occurs because reactive hydroxyl groups are blocked. It follows that increased glycosylation, as in monoglucosides and diglucosides, increases stability.

In an aqueous medium, including foods, anthocyanins can exist in four possible structural forms depending on pH (Fig. 8.11): the blue quinonoidal base (A), the red flavylium cation (AH^+), the colorless carbinol pseudobase (B), and the colorless chalcone (C). Fig. 8.12 shows the equilibrium distributions of these four forms in the pH

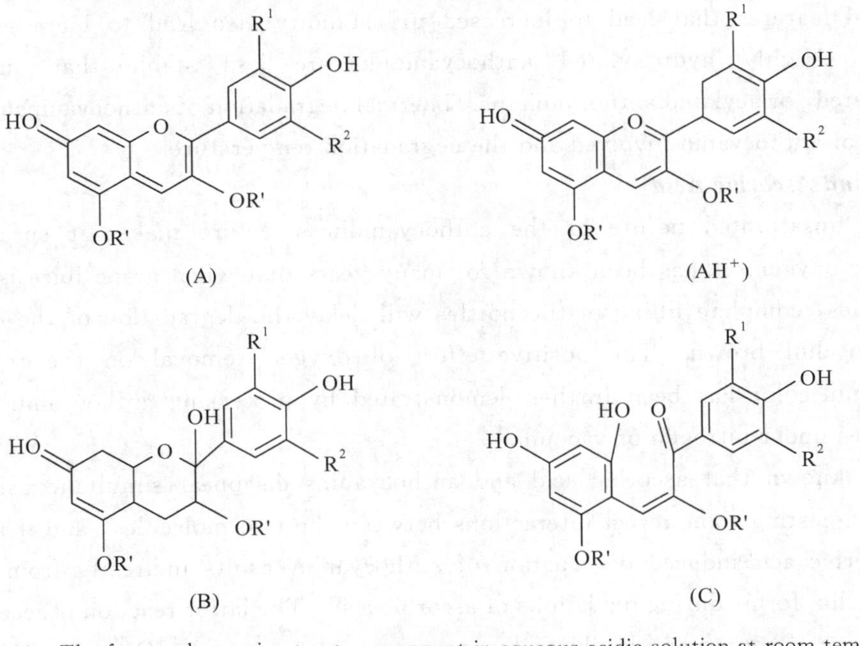

Fig. 8.11 The four anthocyanin structures present in aqueous acidic solution at room temperatures: (A) quinonoidal base (blue); (AH^+) flavylium salt (red); (B) pseudobase or carbinol (colorless); (C) chalcone (colorless)

range 0-6 for malvidin-3-glucoside. It is demonstrated that only two of the four species are important over this pH range. In a solution of malvidin-3-glucoside at low pH the flavylium structure dominates, while at pH 4-6 the colorless carbinol dominates. This solution therefore is colored throughout the 0-6 pH range, turning from red to blue as pH is increased in this range.

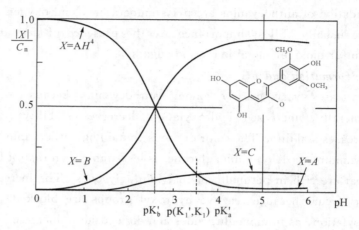

Fig. 8.12 Equilibrium distribution at 25°C of (AH^+, A, B, and C as a function of pH for malvidin 3-glucoside

Temperature

Anthocyanin stability in foods is greatly affected by temperature. In general, structural features that lead to increased pH stability also lead to increased thermal stability. Highly hydroxylated anthocyanidins are less stable than methylated, glycosylated, or acylated anthocyanidins. Thermal degradation of anthocyanins depends on the type of anthocyanin involved and the degradation temperature.

Oxygen and Ascorbic Acid

The unsaturated nature of the anthocyanidin structure makes it susceptible to molecular oxygen. It has been known for many years that when grape juice is hot-filled into bottles, complete filling of the bottles will delay the degradation of the color from purple to dull brown. The positive effect of oxygen removal on the retention of anthocyanin color has been further demonstrated by processing anthocyanin-pigmented fruit juices under nitrogen or vacuum.

It is known that ascorbic acid and anthocyanins disappear simultaneously in fruit juices, suggesting some direct interactions between the two molecules, and it is believed that ascorbic acid-induced degradation of anthocyanin results indirectly from hydrogen peroxide that forms during oxidations of ascorbic acid. The latter reaction is accelerated by the presence of copper and inhibited by the presence of flavonols. Conditions that do not favor formation of H_2O_2 during the oxidation of ascorbic acid therefore account for anthocyanin stability in some fruit juices.

Light

It is generally recognized that light accelerates degradations of anthocyanins. Other forms of radiant energy such as ionizing radiations can also result in anthocyanin degradation.

Sugars and Their Degradation Products

Sugars at high concentrations, as in fruit preserves, stabilize anthocyanins. This effect is believed to result from a lowering of water activity. When sugars are present at concentrations sufficiently low to have little effect on Aw, they or their degradation products sometimes can accelerate the anthocyanin degradation. At low concentrations, fructose, arabinose, lactose, and sorbose have a greater degradative effect on anthocyanins than do glucose, sucrose, and maltose. The reaction is very temperature dependent, and is hastened by the presence of oxygen.

Metals

Metal complexes of anthocyanin are common in the plant world and they extend the color spectrum of flowers. Coated metal cans have long been found to be essential for retaining typical colors of anthocyanins of fruits and vegetables during sterilization in metal cans. Anthocyanins with vicinal, phenolic hydroxyl groups can sequester several multivalent metals. Complexation produces a bathochromic shift toward the blue. Some studies have shown that the metal complexation stabilizes the color of anthocyanin-containing foods.

A fruit discoloration problem referred to as "pinking" has been attributed to the formation of metal anthocynin complexes. This type of discoloration has been reported in pears, peaches, and lychees. It is generally believed that pinking is caused by the heat induced conversion of colorless proanthocyanidins to anthocyanins under acid conditions, followed by the complex formation with metals.

Sulfur Dioxide

One step in the production of maraschino involves bleaching of anthocyanins by SO_2 at high concentrations (0.8%-1.5%). The bleaching effect can be reversible or irreversible. An example of reversible bleaching occurs when fruits containing anthocyanins are preserved against microbial spoilages by holding them in a solution containing 500-2,000 ppm SO_2. During storages the fruit loses its color, but the color can be restored by "desulfuring" (thorough washing) before further processings. In the reversible reaction a colorless complex is first formed. This reaction involves attachment of SO_2 at position C-4 to form the colorless complex shown in Fig. 8.13. The rate of the discoloration reaction is very high, that means a small amount of SO_2 can quickly decolorize a significant amount of anthocyanin. Anthocyanins that are resistant to SO_2 bleaching either have the C-4 position blocked or exist as dimers linked through their 4 position.

Copigmentation

Anthocyanins are known to condense with themselves (self-association) and other organic compounds (copigmentation). Weak complexes form with proteins, tannins, other flavonoids, and polysaccharides. Although most of these compounds themselves are not colored, they augment the color of anthocyanins by causing bathochromic shifts and increased light absorptions at the wavelength of maximum light absorption. These complexes also tend to be more stable during processings and storages. The stable color of wine is believed to result from the self-association of anthocyanin. Such polymers are less pH sensitive and, because the association occurs through the 4 position, are resistant to discoloration by SO_2.

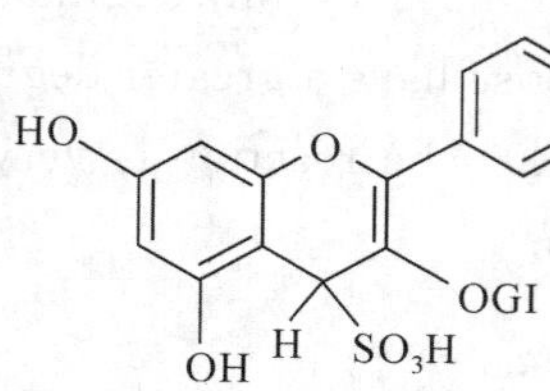

Fig. 8.13 Colorless anthocyanin-sulfate ($-SO_2$) complex

Enzyme Reactions

Enzymes have been involved in the decolorization of anthocyanins. Two groups have been identified: glycosidases and polyphenol oxidases. Together they are generally referred to as anthocyanases. Glycosidases, hydrolyze glycosidic linkages, yielding sugars and the aglycone. Loss of color intensity results from the decreased solubility of the anthocyanidins and their transformations to colorless products. Polyphenol oxidases act in the presence of o-diphenols and oxygen to oxidize anthocyanins. The enzyme first oxidizes the o-diphenol to o-benzoquinone, which in turn reacts with the anthocyanins by a nonenzymatic mechanism to form oxidized anthocyanins and degradation products.

Although blanching of fruits is not a general practice, anthocyanin destroying enzymes can be inactivated by a short blanch treatment (45-60 sec at 90-100 ℃). This has been suggested for sour cherries before the freezing.

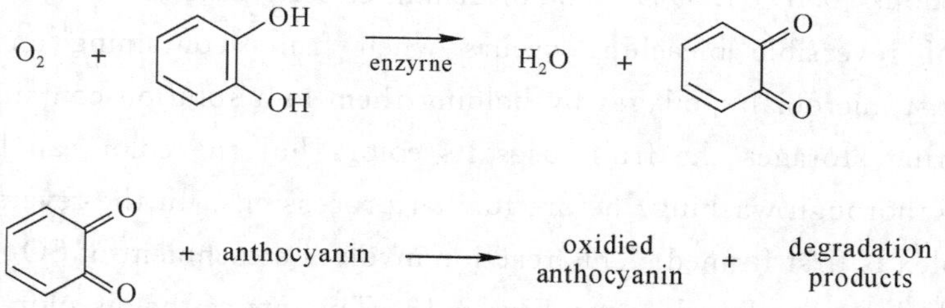

Fig. 8.14 Proposed mechanism of anthocyanin degradation by polyphenol xidase

8.2.4.2 Other Flavonoids

Anthocyanins are the most prevalent flavonoids. Although most yellow colors in food are attributable to the presence of carotenoids, some are attributable to the presence of nonanthocyanin-type flavonoids. In addition, flavonoids also account for some of the whiteness of plant materials, and the oxidation products of those containing phenolic groups contribute to the browns and blacks found in nature. Differences among classes of flavonoids relate to the state of oxidation of the 3-carbon link (Fig. 8.9). Structures commonly found in nature vary from flavan-3-ols (catechin) to flavonols (3-hydroxyflavones) and anthocyanins. The flavonoids also include flavanone, flavononols or dihydroflavonol, and flavan-3,4-diols (proanthocyanidin). In addition, there are five classes of compounds that do not possess the basic flavonoid skeleton, but are chemically related, and therefore are generally included in the flavonoid group. These are the dihydrochalcones, chalcones, isoflavones, neoflavones, and aurones. Individual compounds within this group are distinguished, as with anthocyanins, by the number of hydroxyl, methoxyl, and other substituents on the two benzene rings. Many flavonoid compounds carry a name related to the first source from which they were isolated, rather than being named according to the substituents of the respective aglycone.

Proanthocyanidins

Proanthocyanidins have structural similarities with anthocyanidins. They can be converted to colored products during food processing. Proanthocyanidins are also referred to as leucoanthocyanidins or leucoanthocyanins. The term leucoanthocyanidin is appropriate if it is used to designate the monomeric flavan-3,4-diol (Fig. 8.15), which is the basic building unit of proanthocyanidins. The latter can occur as dimers, trimers, or higher polymers. The intermonomer linkage is generally through carbons 4→8 or 4→6.

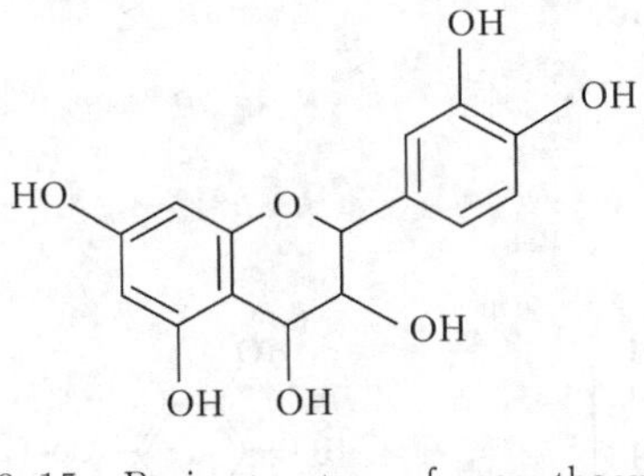

Fig. 8.15 Basic structure of proanthocyanidin

Proanthocyanidins were first found in cocoa beans, where upon heating under acidic conditions they hydrolyze into cyanidin and epicatechin (Fig. 8.16). Dimeric proanthocyanidins have been found in apples, pears, kola nuts, and other fruits. These compounds are known to degrade in air or under light to red-brown stable derivatives. They contribute significantly to the color of apple juice and other fruit juices, and to astringency in some foods. To produce astringency, proanthocyanidins of two to eight units interact with proteins. Other proanthocyanidins found in nature will yield on hydrolysis the common anthocyanidins—pelargonidin, petunidin, or delphinidin.

8.2.4.3 Tannins

Tannins are special phenolic compounds and are given this name simply by virtue of

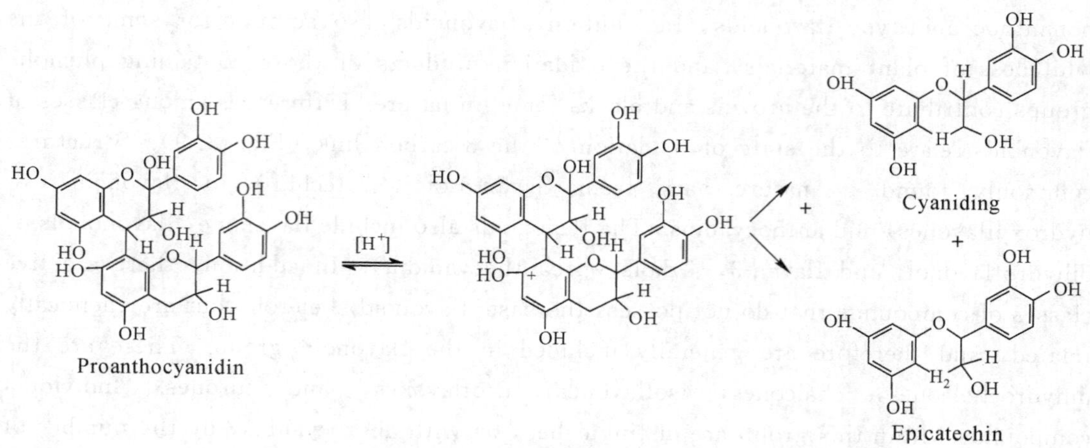

FOOD CHEMISTRY

Fig. 8.16 Mechanism of acid hydrolysis of proanthocyanidin.

their ability to combine with proteins and other polymers such as polysaccharides, rather than their exact chemical natures. They are functionally defined, therefore, as water-soluble polyphenolic compounds with molecular weights between 500 and 3,000 that have the ability to precipitate alkaloids, gelatin, and other proteins. They occur in the bark of oak trees and in fruits and they are generally considered as two groups: (i) proanthocyanidins, also referred to as "condensed tannins", and (ii) glucose polyesters of gallic acid of hexahydsoxydiphenic acids (Fig. 8.17). The latter group is also known as hydrolyzable tannins, because they consist of a glucosidic molecule bonded to different phenolic moities. The most important example is glucose bonded to gallic acid and the lactone of its dimer, ellagic acid.

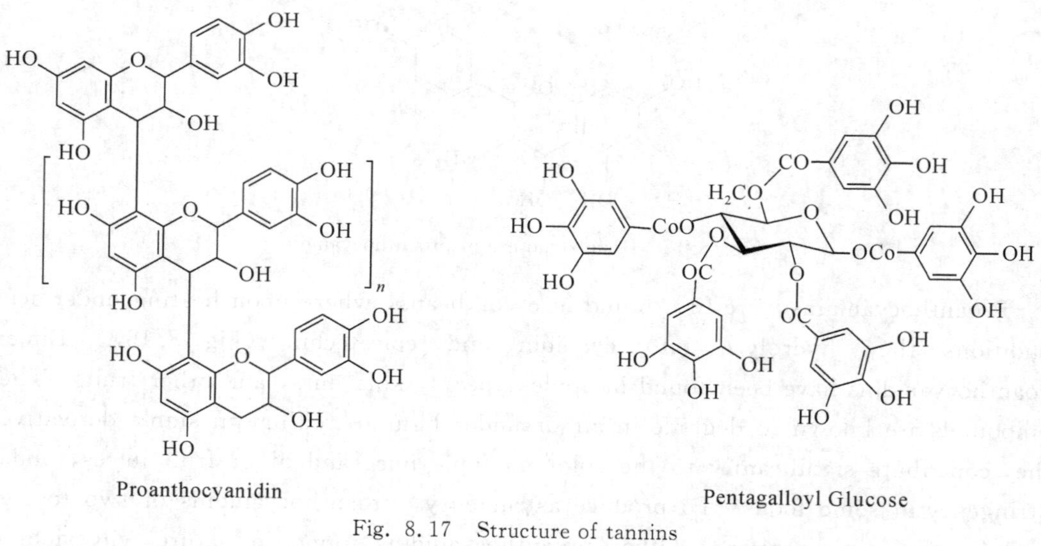

Fig. 8.17 Structure of tannins

Tannins range in color from yellowish-white to light brown and contribute to astringency in foods. Their ability to precipitate proteins makes them valuable as clarifying agents.

8.3 Synthetic Colorants

Because of their bright stable hue, the synthetic colorants are usually applied in foods in addition to natural pigments. To compare with the natural pigments, consumers concerned more about the safety of artificial-synthesized colorants. Hence all the food used synthetic colorants are legislatively authorized by the governments or international organizations. Currently permitted synthetic colorants by some nations and international organizations are listed in Table 8.3.

Colorants exempt from certification are natural pigments or specific synthetic dyes that are nature identical. An example of the latter is -carotene, which is widely distributed in nature but also can be synthesized to achieve a "nature identical" substance. Color additives currently exempt form certifications are listed in Table 8.4.

FAO and WHO have devised the "acceptable daily intakes" (ADI) for food additives, including colorants (Table 8.5). Those additives for which an ADI value has been established and approved, including six artificial dyes, are considered not to present a hazard to consumers. Those additives for which the safety evaluation is not complete, and therefore these additives have a provisional use status. Included in this category are colorants derived from beet, annatto, and turmeric, all of which are approved for use in the United States.

Table 8.3 Synthetic color additives currently permitted by the nations and international organization

Name	EEC number	FDA number	EEC	United States	Canada	Japan
Erythrosine	E123	FD&C Red No. 3	+	+	+	+
Brilliant blue FCF	—	FD&C Blue No. 1	$-^c$	+	+	+
Indigoting	E132	FD&C Blue No. 2	+	+	+	+
Tartrazine	E102	FD&C Yellow No. 5	$+^d$	+	+	+
Quinoline yellow	E104	FD&C Yellow No. 6	$+^e$	—	—	—
Alhura red	—	FD&C Red No. 40	—	+	+	—
Yellow 2G	E107	—	$+^f$	—	—	—
Ponceau 4R	E124	—	$+^d$	—	—	+
Camoisine	E122	—	$+^{d,e,g}$	—	—	—
Amaranth	E123	FD&C Red No. 2	$+^d$	—	+	—
Red 2G	E128	—	$+^f$	—	—	—
Patent blue	E131	—	$+^e$	—	—	—
Green S	E142	—	$+^{d,e,g}$	—	—	—

(To be continued)

FOOD CHEMISTRY

(Table 8.3)

Name	EEC number	FDA number	EEC	United States	Canada	Japan
Brown FK	E154	—	$+^f$	—	—	—
Chocolate brown HT	E155	—	$+^b$	—	—	—
Black PN	E151	—	$+4^e$	—	—	—

[a] +, Permitted for food use, (in some countries limited to specific foods); -, prohibited for food use.
[b] No synthetic colorants permitted in Norway.
[c] Permitted in Denmark, Ireland, and Netherlands.
[d] Not permitted in Finland.
[e] Not permitted in Portug al.
[f] Permitted in Ireland only.
[g] Not permitted in Sweden.
[h] Permitted in Ireland and Netherlands.

Table 8.4 U.S. color additives currently exempt from certification

Color	Use limitation[a]
Annatto extract	—
Dehydrated beet (beet powder)	—
Canthaxanthin	Not to exceed 66 mg/kg of solid or pint of liquid food
β-Apo-S'-carotenal	Not to exceed 15 mg/lb or liter of food
β-Carotene	—
Caramel	33mg/kg
Cochineal extract; carmine	—
Toasted, partially defatted, cooked cottonseed flour	—
Ferrous gluconate	Colorant for ripe olives
Crape-skin extract (enocianina)	Colorant for beverages only
Synthetic iron oxide	Not to exceed 0.25% by weight of pet food
Fruit juice	—
Vegetable juice	—
Dried algae metal	Chicken feed only; to enhance yellow olor of skin and eggs
Tagetes (Aztec marigold) and extract	Chicken feed only; to enhance yellow color of skin and eggs
Carrot oil	—
Corn endospem oil	Chicken feed only; to enhance yellow color of skin and eggs
Paprika	—
Paprika oleoresin	—
Riboflavin	—
Saffron	—
Titanium dioxide	Not to exceed 1% by weight of food
Tumeric	—
Tumeric oleoresm	—

[a] Unless otherwise stated the colorant may be used in an amount consistent with good manufacturing practices, except for foods where use is specified by a standard of identity.

Table 8.5 Acceptable daily intake of some synthetic and natural colorants

Synthetic color	Intake[a] (mg/kg)	Natural color	Amount (mg/kg)
Tartrazine	7.5	β-Apo-8'-carotenal	2.5
Sunset yellow FCF	5.0	β-Carotene	5.0
Amaranth	1.5	Canathaxanthin	25.0
Erythrosine	1.25	Riboflavin	0.5
Brilliant blue FCF	12.5	Chlorophyll	GMP[b]
Indigotine	2.5	Caramel	GMP

[a] Daily intake over lifetime without risks.
[b] Amount consistent with "good manufacturing practice".

The chemical structure of the seven synthetic colorants are shown in Fig. 8.18.

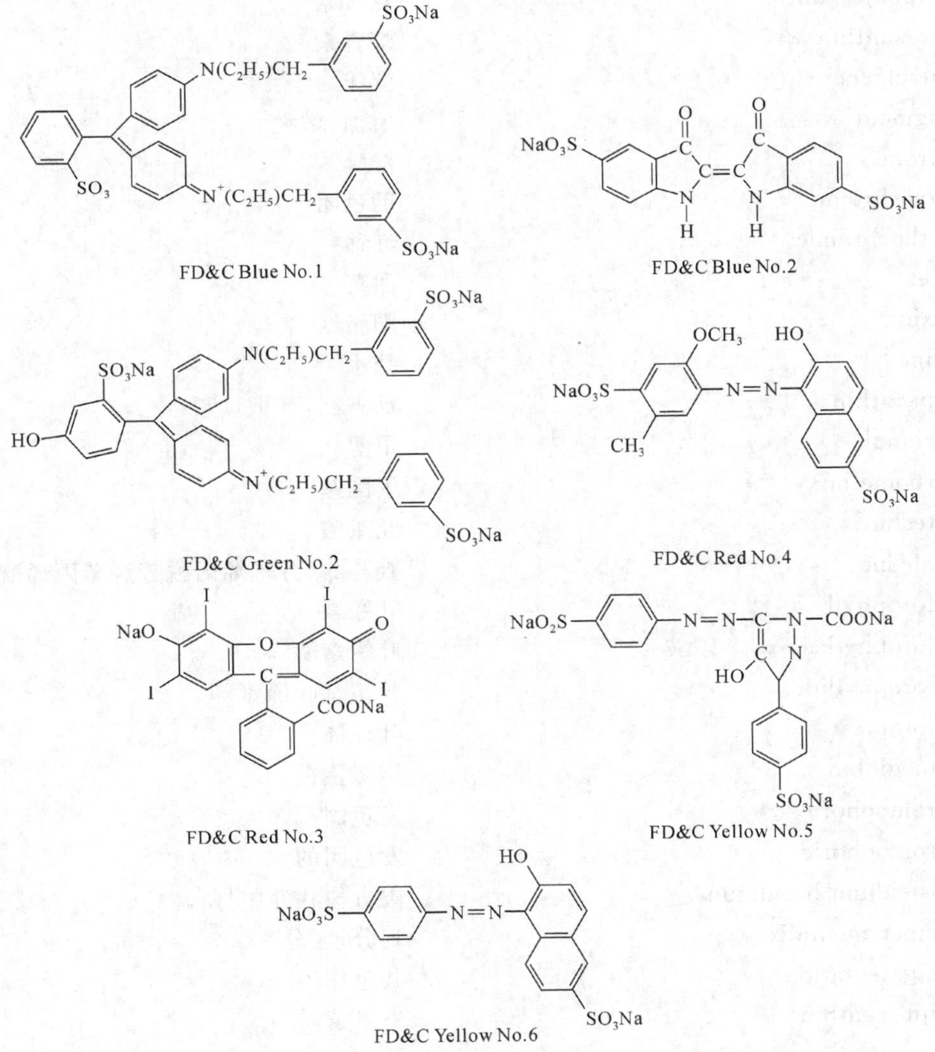

Fig. 8.18 Structures of certified color additives currently permitted for the general use in the United States

Glossary

acetone	丙酮
aglycone	糖苷配基
algae	藻类,海藻
alkaloid	生物碱
Alumina	氧化铝
annatto	胭脂树,胭脂红
anthocyanidin	花青素
astaxanthin	虾青素
astringency	涩味
augment	增加,增大
aurone	橙酮
auxochrome	助色团
bathochromic	红移
beet	甜菜
bixin	胭脂红
brine	盐水
capsanthin	辣椒红,辣椒黄素
carbinol	甲醇
carbomethoxy	甲酯基
catechin	儿茶酸
chalcone	查耳酮(芳基烯丙酰芳烃苯丙烯酰苯)
chlorophyll	叶绿素
chlorophyllase	叶绿素酶
chlorophyllide	脱植基叶绿素
chloroplast	叶绿体
choleglobin	胆绿蛋白
chromophore	发色团
chromophoric	发色团的
clostridium botulinum	肉毒梭状芽胞杆菌
compartmentalize	划分,区分
copigmentation	共色作用
cryptoxanthin	玉米黄质
cured	薰,腌

cyanidin	花青色素
cyanidin	花青色素
delphinidin	飞燕草色素
dihydrochalcone	二氢查耳酮
dihydroflavonol	二氢黄酮醇
dimethylformamide	二甲基亚酰胺
diol	二醇
ellagic acid	鞣花酸
epicatechin	表儿茶素,表儿茶酸
epoxide	环氧化合物
epoxy	环氧基树脂
epoxy-β-ionone	环氧-β-紫罗兰酮
esterase	酯酶
ethyl	乙基,乙烷基
flavan-3-ols	黄烷-3-醇
flavanone	黄酮醇(衍生物)
flavonol	黄酮醇
flavononol	黄烷酮醇
flavylium	即花色基元
formyl	甲酰
gallate	没食子酸盐
gallic acid	没食子酸
gelatin	明胶,凝胶
glycosylation	糖基化
hasten	加快
Heme	亚铁血红素
hemoglobin	血红蛋白
hexane	己烷
hindrance	障碍,阻碍
hydrogen sulfide	硫化氢
hydroxyl	羟基
hydroxylation	羟基化
hypsochromic	蓝移
ionone	紫罗酮
isoflavone	异黄酮,大豆异黄酮
isoprene	异戊二烯
isoprenoid	类异戊二烯
lactone	内酯

lakes	色淀
lamellae	片晶,薄片
leucoanthocyanidin	无色花色素
leucoanthocyanin	无色花色苷
ligand	配基
lime	绿黄色,酸橙色
lobster	龙虾
lutein	叶黄素
lycopene	番茄红素
Lye	碱液
macrocyclic	大环的
malvidin	锦葵色素
maraschino	黑樱桃酒
marigold	万寿菊
methoxy	含甲氧基的
methylation	甲基化
myoglobin	肌红蛋白
neoflavone	新黄酮
neoxanthin	新黄素,新叶黄素
nitric oxide	一氧化氮
nitrihemin	硝化氯化血红素
nitrite	亚硝酸盐
nitrosylhemochrome	亚硝基血色原
nitrosylmyoglobin	亚硝基肌红蛋白
nucleus	原子核,核心
o-benzoquinone	邻醌
okra	秋葵
operative	有效的,运转着的
organelle	细胞器,细胞器官
oxygenated xanthophylls	氧合叶黄素
pelargonidin	花葵素
peroxidation	过氧化反应
petunidin	苷元
pheophorbide	脱酶叶绿素盐
pheophytin	脱酶叶绿素
phytol	叶绿醇,植醇
pigmentation	染色,天然颜色
porphin	卟吩

porphyrins	卟啉
proanthocyanidin	原花色素
prooxidant	促氧化
propionate	丙酸盐,丙酸酯
propyl	丙基
provisional	临时的,暂时的
pseudobase	假碱
pyridine	嘧啶
pyrrole	吡咯
quinonoidal	醌式碱
retina	视网膜
sequester	使隔绝
squash	挤压,南瓜属植物
Steric	空间的,立体的
substituent	取代基
substratum	基础,下层
tetrapyrrole	四吡咯
turmeric	姜黄
vicinal	邻近的
vinyl	乙酰基
violaxanthin	黄质,紫黄素
zeaxanthin	玉米黄质

Chapter 9 Flavors

9.1 Introduction

Generally, the term "flavor" has evolved to a usage that implies an overall integrated perception of all of the contributing senses (smell, taste, sight, feeling, and sound) at the time of food consumptions. The ability of specialized cells of the olfactory epithelium of the nasal cavity is to detect trace amounts of volatile odorants accounts for the nearly unlimited variations in intensity and quality of odors and flavors. Taste buds located on the tongue and back of the oral cavity enable humans to sense sweetness, sourness, saltiness, and bitterness, and these sensations contribute to the taste component of flavor. Nonspecific or trigeminal neural responses also provide important contributions to flavor perception through the detection of pungency, cooling, umami, or delicious attributes, as well as other chemically induced sensations that are incompletely understood. The nonchemical or indirect senses (sight, sound, and feeling) influence the perception of tastes and smells, and hence food acceptances, but a discussion of these effects is beyond the scope of this chapter. Thus, materials presented in this chapter will be confined to discussions about substances that yield taste and/or odor responses, but a clear distinction between the meaning of these terms and that of flavor will not always be attempted.

9.2 Taste and Nonspecific Saporous Sensations

Frequently, substances responsible for these components of flavor perception are water soluble and relatively nonvolatile. As a general rule, they are also present at higher concentrations in foods than those responsible for aromas, and have been often treated lightly in coverages of flavors. Because of their extremely influential role in the acceptance of food flavors, it is appropriate to examine the chemistry of substances responsible for taste sensations as well as those responsible for some of the less defined flavor-taste sensations.

9.2.1 Taste Substances: Sweet, Bitter, Sour, and Salty

Sweet substances have been the focus of much attention because of interest in sugar alternatives and the desire to find suitable replacements for the low-calorie sweeteners saccharin and cyclamate. The bitterness sensation appears to be closely related to sweetness from a molecular structure-receptor relationship, and as a result much has been learned about bitterness in studies directed primarily toward sweetness. Development of bitterness in protein hydrolysates and aged cheeses is a troublesome problem, and this problem has stimulated research on the causes of bitterness in peptides. With regard to saltiness, national policies that encourage a reduction of sodium in diets have stimulated renewed interest in the mechanisms of the salty taste.

9.2.1.1 Structural Basis of the Sweet Modality

Before modern sweetness theories were advanced, it was popular to deduce that sweetness was associated with hydroxyl ($-OH$) groups because sugar molecules are dominated by this feature.

However, this view was soon subject to criticism because polyhydroxy compounds vary greatly in sweetness, and many amino acids, some metallic salts, and unrelated compounds, such as chloroform ($CHCl_3$) and saccharin, are also sweet. Still, it was apparent that some common characteristics existed among sweet substances, and over the past 75 years a theory relating molecular structures and sweet tastes has satisfactorily explained why certain compounds exhibit sweetness.

9.2.1.2 The Bitter Taste Modality

Bitterness resembles sweetness because of its dependence on the stereochemistry of stimulus molecules, and the two sensations are triggered by similar molecule features. Although sweet molecules must contain two polar groups that may be supplemented with a nonpolar group, bitter molecules appear to have a requirement for only one polar group and a hydrophobic group. However, some believe that most bitter substances possess an AH/B entity identical to that found in sweet molecules as well as the hydrophobic group.

9.2.1.3 The Salty and Sour Taste Modalities

Classic salty taste is represented by sodium chloride (NaCl), and is also given by lithium chloride (LiCl). Salts have complex tastes, consisting of psychological mixtures of sweet, bitter, sour, and salty perceptual components. Furthermore, it has been shown recently that the tastes of salts often fall outside the traditional taste sensation and are difficult to describe in classic terms. Nonspecific terms, such as chemical and soapy, often seem to more accurately describe the sensations produced by salts than do the classic terms. Chemically, it appears that cations cause salty tastes and anions modify salty tastes.

Sodium and lithium produce only tastes, while potassium and other alkaline earth cations produce both salty and bitter tastes. Among the anions commonly found in foods, the chloride ion is least inhibitory to the salty taste, and the citrate anion is more inhibitory than orthophosphate anions. Anions not only inhibit tastes of cations, but also contribute tastes themselves. The chloride ion does not contribute a taste, and the citrate anion contributes less than the orthophosphate anion. Anion taste effects impact the flavor of foods.

The most acceptable model for describing the mechanism of salty taste perception involves the interaction of hydrated cationanion complexes with AH/B-type receptor sites (discussed earlier). The individual structures of such complexes vary substantially, so that both water OH groups and salt anions or cations associate with receptor sites.

Similarly, the perception of sour compounds is convinced to involve an AH/B-type receptor site. However, datas are not sufficient to determine whether hydronium ions (H_3O^+), dissociated inorganic or organic anions, or nondissociated molecular species are most influential in the sour response. Contrary to popular beliefs, the acid strength in a solution does not appear to be the major determinant of the sour sensation; rather, other poorly understood molecular features appear to be of the primary importance (e. g., molecular weight, size, and overall polarity).

9.2.2 Flavor Enhancers

Compounds eliciting this unique effect have been utilized by humans since the inception of the food cooking and preparation, but the actual mechanism of flavor enhancement remains largely a mystery. These substances contribute a delicious or umami taste to foods when used at levels in excess of their independent detection threshold, and they simply enhance flavors at levels below their independent detection thresholds. Their effects are prominent and desirable in the flavors of vegetables, dairy products, meats, poultry, fish, and other seafoods. The best known members of this group of substances are the 5'-ribonucleotides, of which 5'-inosine monophosphate or 5'-IMP (XI) serves as a suitable example, and monosodium L-glutamate (MSG) (XII).

(XI) 5'-Inosine monophosphate

(XII) L-Glutamate, Na^+ (MSG)

It is well documented that a synergistic interaction occurs between MSG and the 5'-ribonucleotides in both providing the umami taste and in enhancing flavors. This suggests that some common structural features exist among active compounds. Although most attention has been directed toward the 5'-ribonucleotides and MSG, other flavor-enhancing compounds have been claimed to exist. Maltol and ethyl maltol are worthy of mention because they are used commercially as flavor enhancers for sweet goods and fruits. At high concentrations maltol possesses a pleasant, burnt caramel aroma. It provides a smooth, velvety sensation to fruit juices when it is employed in concentrations of about 50 ppm.

Maltol and ethyl maltol ($-C_2H_5$ instead of $-CH_3$ on the ring) both could fit the AH/B portion of the sweet receptor, but ethyl maltol is more effective as a sweetness enhancer than maltol. Maltol lowers the detection threshold concentration for sucrose by a factor of two. The actual mechanism for the enhancing effects of these compounds is unknown. Similar compounds of this type are derived naturally from browning reactions and are noted later in the chapter in the section on the development of "reaction" flavors.

9.2.3 Astringency

Astringency is a taste-related phenomenon, perceived as a dry feeling in the mouth along with a coarse puckering of the oral tissue. Astringency usually involves the association of tannins or polyphenols with proteins in the saliva to form precipitates or aggregates. Additionally, sparingly soluble proteins such as those found in certain dry milk powders also combine with proteins and mucopolysaccharides of saliva and cause astringency. Astringency is often confused with bitterness because many individuals do not clearly differentiate its nature, and many polyphenols or tannins cause both astringent and bitter sensations.

Tannins (Fig. 9.1) have broad cross-sectional areas and one suitable for hydrophobic associations with proteins. They also contain many phenolic groups that can convert to quinoid structures; these in turn can cross-link chemically with proteins. Such cross-links

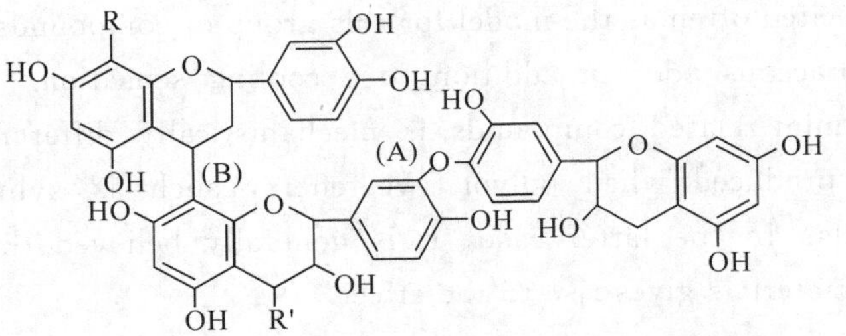

Fig. 9.1 Structure of a procyanidin-type tannin showing condensed tannin linkage (B) and hydrolyzable tannin linkage (A) and also showing large hydrophobic areas capable of associating with proteins to cause astringency

have been suggested as a possible contributor to astringency activity.

Astringency may be a desirable flavor property, such as in tea. However, the practice of adding milk or cream to tea removes astringency through binding of polyphenols with milk proteins. Red wine is a good example of a beverage that exhibits both astringency and bitterness caused by polyphenols. Astringency derived from polyphenols in unripe bananas can also lead to an undesirable taste in products to which the bananas have been added.

9.2.4 Pungency

Certain compounds found in several spices and vegetables cause characteristic hot, sharp, and stinging sensations that are known collectively as pungency. Although these sensations are difficult to separate from those of general chemical irritation and lachrymatory effects, they are usually considered as separate flavor-related sensations. Some pungent principles, such as those found in chili pepper, black pepper, and ginger, are not volatile and exert their effects on oral tissues. Other spices and vegetables contain pungent principles that are somewhat volatile, and produce both pungency and characteristic aromas. These include mustard, horseradish, vegetable radishes, onions, garlic, watercress, and the aromatic spice, clove, which contains eugenol as the active component. All these spices and vegetables are used in foods to provide characteristic flavors or to generally enhance the palatability. Usage at low concentrations in processed foods frequently provides liveliness to flavors through subtle contributions that fill out the perceived flavors.

9.2.5 Cooling

Cooling sensations occur when certain chemicals contact the nasal or oral tissues and stimulate a specific saporous receptor. These effects are most commonly associated with mint-like flavors, including peppermint, spearmint, and wintergreen. Several compounds cause the sensation but (−)-menthol (XVI), in the natural form (1-isomer), is most commonly used in flavors. The fundamental mechanism of the cooling sensation is not known, but compounds yielding this sensation also produce accompanying aromas. Camphor (XVII) is cited often as the model for this group of compounds, and it produces a distinctive camphoraceous odor in addition to a cooling sensation. The cooling effect produced by the mint-related compounds is mechanistically different from the slight cooling sensation produced when polyol sweeteners, such as xylitol, are tasted as crystalline materials. In the latter case, it is generally believed that an endothermic dissolution of the materials gives rise to the effect.

(XVI) (−)-Menthol

(XVII) d-Camphor

9.3 Vegetable, Fruit, and Spice Flavors

Categorization of vegetable and fruit flavors in a reasonably small number of distinctive groups is not easy, since logical groupings are not necessarily available for vegetables and fruits. For example, some information on plant-derived flavors was presented in the section on pungency, and some are covered in the section dealing with the development of "reaction" flavors. Emphasis in this section is on the biogenesis and development of flavors in important vegetables and fruits. For information on the other fruit and vegetable flavors, the reader is directed to the general references.

9.3.1 Sulfur-Containing Volatiles in Allium *sp*.

Plants in the genus Allium are characterized by strong, penetrating aromas, and important members are onions, garlic, leek, chives, and shallots. These plants lack the strong characterizing aroma unless the tissue is damaged and enzymes are decompartmentalized so that flavor precursors can be converted to odorous volatiles. In the case of onions (*A. cepa L.*), the precursor of the sulfur compounds that are responsible for the flavor and aroma of this vegetable is S-(1-propenyl)-L-cysteine sulfoxide. This precursor is also found in leeks. Rapid hydrolysis of the precursor by allinase yields a hypothetical sulfenic acid intermediate along with ammonia and pyruvate (Fig. 9.2). The sulfenic acid undergoes further rearrangements to yield the lachrymator, thiopropanal S-oxide, which is associated with the overall aroma of fresh onions. The pyruvic acid produced by the enzymatic conversion of the precursor compound is a stable product of the reaction and serves as a good index of the flavor intensity of onion products. Part of the unstable sulfenic acid also rearranges and decomposes to a rather large number of compounds in the classes of mercaptans, disulfides, trisulfides, and thiophenes. These compounds also comprise the flavor substances which provide cooked onion flavors. The flavor of garlic (Allium sativum L.) is formed by the same general type of mechanism that functions in onion, except that the precursor is S-(2-propenyl)-L-cysteine sulfoxide. Diallyl thiosulfinate (allicin) (Fig. 9.3) contributes to flavor of garlic, and an S-oxide lachyramator similar to that formed in onion is not formed. The thiosulfinate flavor compound of garlic decomposes and rearranges in much the same manner as indicated for the sulfenic acid of onion (Fig. 9.2). This results in methyl allyl and diallyl disulfides and other principles in garlic oil and cooked garlic flavors.

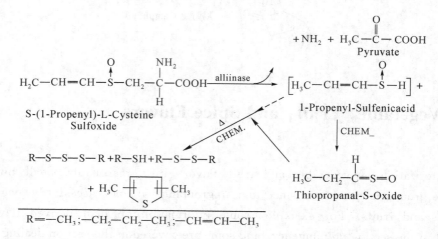

Fig. 9.2 Reactions involved in the formation of onion flavor (chem. = nonenzymic)

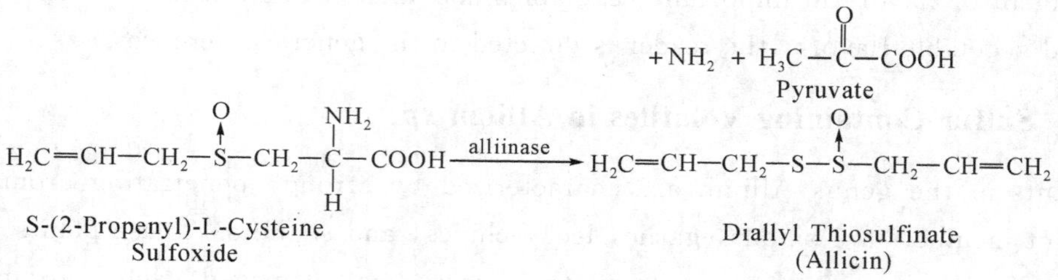

Fig. 9.3 Formation of the principal fresh garlic flavor compounds

9.3.2 Sulfur-Containing Volatiles in the Cruciferae

The Cruciferae family contains Brassica plants such as cabbage (Brassica oleracea capitata L.), brussel sprouts (Brassica oleracea var. gemmifera L.), turnips (Brassica rapa var. rapa L.), and brown mustard (Brassica juncea Coss.), as well as watercress (Nasturtium officinale R. Br.), radishes (Raphanus sativus L.), and horseradish (Armoracia lapathifolia Gilib). As noted in the discussion about pungent compounds, the active pungent principles in the Cruciferae are also volatile and therefore contribute to characteristic aromas. Further, the pungency sensation frequently involves irritation sensations, particularly in the nasal cavity, and lachrymatory effects. The flavor compounds in these plants are formed through enzymic processes in disrupted tissues and through cooking. The fresh flavors of the disrupted tissue are caused mainly by isothiocyanates resulting from the action of glucosinolases on thioglycoside precursors. The reaction shown in Fig. 9.4, yielding allyl isothiocyanate, is illustrative of the flavor-forming mechanism in Cruciferae, and the resulting compound is the main source of pungency and aroma in horseradish and black mustard.

Cyanates resulted from the action of glucosinolases on thioglycoside precursors. The reaction shown in Fig. 9.4, yielding allyl isothiocyanate, is illustrative of the flavor-forming mechanism in Cruciferae, and the resulting compound is the main source of pungency and aroma in horseradish and black mustard.

Several glucosinolates (S-glycosides) occur in the cruciferae, and each gives rise to characteristic flavors. The mild pungency or radishes is caused by the aroma compound 4-methylthio-3-t-butenylisothiocyanate (XVIII). In addition to the isothiocyanates, glucosinolates also yield thiocyanates (R—S=C=N) and nitriles (Fig. 9.4).

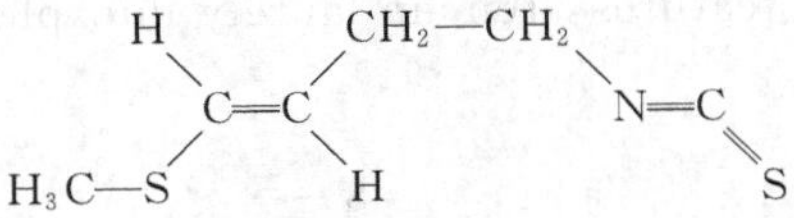

(XVIII 4-Methyithio-3-t-Butenyisothiocyanate)

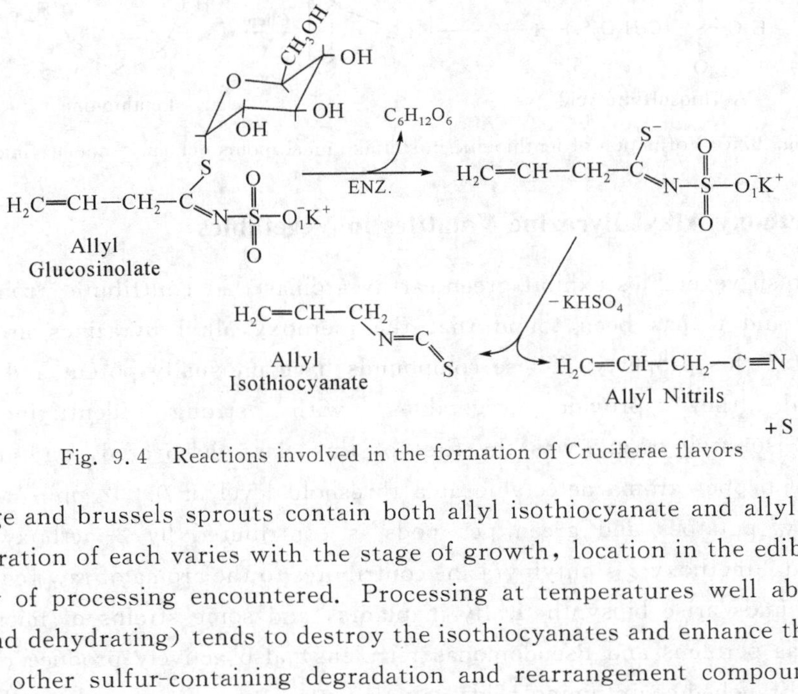

Fig. 9.4 Reactions involved in the formation of Cruciferae flavors

Cabbage and brussels sprouts contain both allyl isothiocyanate and allyl nitrile, and the concentration of each varies with the stage of growth, location in the edible part, and the severity of processing encountered. Processing at temperatures well above ambient (cooking and dehydrating) tends to destroy the isothiocyanates and enhance the amount of nitriles and other sulfur-containing degradation and rearrangement compounds. Several aromatic isothiocyanates occur in Cruciferae; for example, 2-phenylethyl isothiocyanate is one of the main aroma compounds of watercress. This compound also contributes a tingling pungent sensation.

9.3.3 Unique Sulfur Compound in Shiitake Mushrooms

A novel C-S lyase enzyme system has been discovered in Shiitake mushrooms (lentinus edodes), which are prized in Japan for their delicious flavors. The precursor for the major flavor contributor, lentinic acid, is an S-substituted L-cysteine sulfoxide bound as a g-glutamyl peptide. The initial enzyme reaction in the flavor development involves a g-glutamyl transpeptidase, which releases a cysteine sulfoxide precursor (lentinic acid). Lentinic acid is then attacked by S-alkyl-L-cysteine sulfoxide lyase (Fig. 9.5) to yield to lenthionine, an active flavor compound. These reactions are initiated only after the tissue is disrupted, and the flavor develops only after drying and rehydration or after holding freshly mascerated tissue for a short period of time. Other polythiepanes in addition to lenthionine are formed, but the flavor is ascribed to lenthionine. Flavor roles for other mixed-atom polysulfides with related structural features, remain largely unexplored.

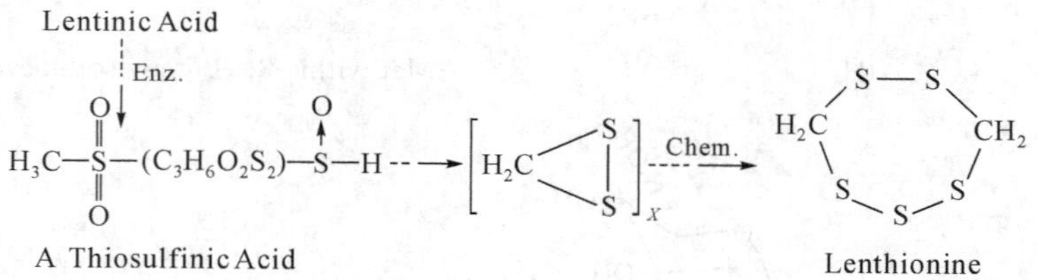

Fig. 9.5 Formation of lenthionine in shiitake mushrooms (chem. =nonenzymic)

9.3.4 Methoxy Alkyl Pyrazine Volatiles in Vegetables

Many fresh vegetables exhibit green-earthy aromas that contribute strongly to their recognition, and it has been found that the methoxy alkyl pyrazines are frequently responsible for this property. These compounds have unusually potent and penetrating odors, and they provide vegetables with strong identifying aromas. 2-Methoxy-3-isobutylpyrazine was the first of this class discovered, and it exhibits a powerful bell pepper aroma detectable at a threshold level of 0.002 ppb. Much of the aroma of raw potatoes and green pea pods is contributed by 2-methoxy-3-isopropyl pyrazine, and 2-methoxy-3-s-butylpyrazine contributes to the aroma of raw red beet roots. These compounds arise biosynthetically in plants, and some strains of microorganisms (Pseudomonas perolens and Pseudomonas tetrolens) also actively produce these unique substances. Branched-chain amino acids serve as precursors for methoxy alkyl pyrazine volatiles, and the mechanistic scheme shown in Fig. 9.6 has been proposed.

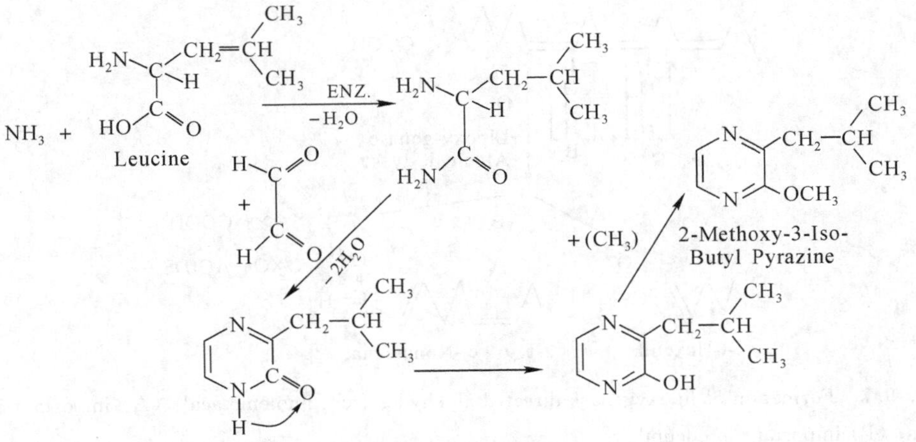

Fig. 9.6 Proposed enzymatic scheme for the formation of methoxy alkyl pyrazines

9.3.5 Enzymically Derived Volatiles from Fatty Acids

Enzymically generated compounds derived from long-chain fatty acids play an extremely important role in the characteristic flavors of fruits and vegetables. In addition, these types of reactions can lead to important off flavors, such as those associated with processed soybean proteins. Further information about these reactions can be found in the discussions about lipids and enzymes.

9.3.5.1 Lipoxygenase-Derived Flavors in Plants

In plant tissues, enzyme-induced oxidative breakdown of unsaturated fatty acids occurs extensively, and this yields characteristic aromas associated with some ripening fruits and disrupted tissues. In contrast to the random production of lipid-derived flavor compounds by purely autoxidizing systems, very distinctive flavors occur when the compounds produced are enzyme determined. The specificity for flavor compounds is illustrated in Fig. 9.7, where the production of 2-t-hexenal and 2-t, 6-c-nonadienal by site-specific hydroperoxidation of a fatty acid is dictated by a lipoxygenase and a subsequent lyase cleavage reaction. Upon cleavage of the fatty acid molecule, oxoacids are also formed, but they do not appear to influence flavors. Decompartmentalization of enzymes is required to initiate this and other reactions, and since successive reactions occur, overall aromas change with time. For example, the lipoxygenase-derived aldehydes and ketones are converted to corresponding alcohols (Fig. 9.8), which usually have higher detection thresholds and heavier aromas than the parent carbonyl compounds. Although not shown, cis-trans isomerases are also present that convert cis-3 bonds to trans-2 isomers. Generally, C_6 compounds yield green plant-like aromas like fresh-cut grass, C_9 compounds smell like cucumbers and melons, and C_8 compounds smell like mushrooms or violet and geranium leaves. The C_6 and C_9 compounds are primary alcohols and aldehydes; the C_8 compounds are secondary alcohols and ketones.

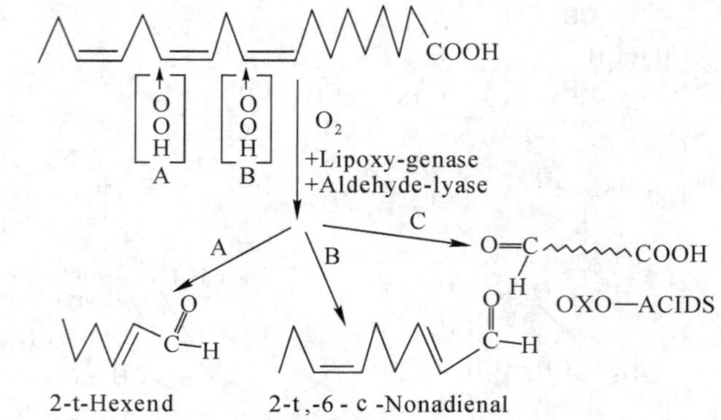

Fig. 9.7 Formation of lipoxygenase-directed aldehydes from linolenic acid: (A) important in fresh tomatoes; (B) important in cucumbers

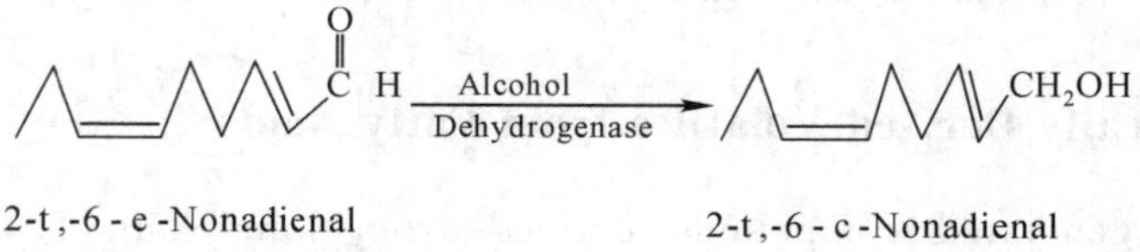

Fig. 9.8 Conversion of aldehyde to alcohol resulting in subtle flavor modification

9.3.5.2 Volatiles from β-Oxidation of Long-Chain Fatty Acids

The development of pleasant, fruity aromas is associated with the ripening of pears, peaches, apricots, and other fruits, and these aromas are frequently dominated by medium chain-length (C_6-C_{12}) volatiles derived from long-chain fatty acids by β-oxidation. The formation of ethyl deca-2-t-4-c-dienoate by this means is illustrated in Fig. 9.9. This ester is the impact or characterizing aroma compound in the Bartlett pear. Although not included in Fig. 9.9, hydroxy acids (C_8-C_{12}) are also formed through this process, and they cyclize to yield g and d-lactones. The C_8-C_{12} lactones possess distinct coconut-like and peach-like aromas characteristic of these respective fruits.

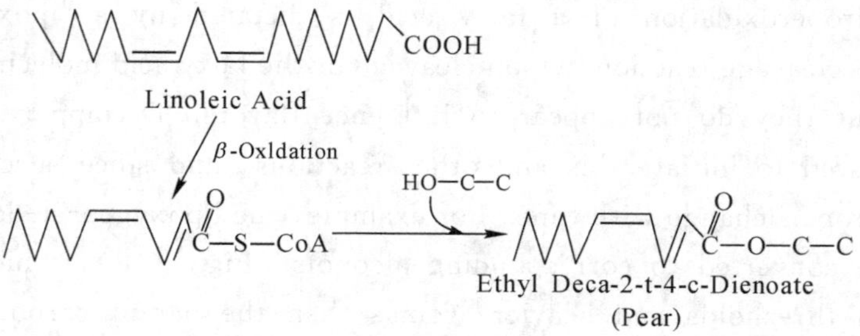

Fig. 9.9 Formation of a key aroma substance in pears throbu-gohxidation of linoleic acid followed by esterification

9.3.6 Volatiles from Branched-Chain Amino Acids

Branched-chain amino acids serve as important flavor precursors for the biosynthesis of compounds associated with some ripening fruits. Bananas and apples are particularly good examples for this process because much of the ripe flavor of each is caused by volatiles from amino acids. The initial reaction involved in flavor formation (Fig. 9.10) is sometimes referred to as enzymic Strecker degradation because the transamination and decarboxylation occur that are parallel to those occuring during the non-enzymatic browning. Several microorganisms, including yeast and malty flavor-producing strains of Streptoccus lactis, can also modify most of the amino acids in a fashion similar to that show in Fig. 9.10. Plants can also produce similar derivatives from amino acids other than leucine, and the occurrence of 2-phenethanol with a rose-or lilac-like aroma in blossoms is attributed to these reactions.

Although the aldehydes, alcohols, and acids from these reactions contribute directly to the flavors of ripening fruits, the esters are the dominant character-impact compounds. It has long been known that isoamyl acetate is important in banana flavor, but other compounds are also required to give full banana flavor. Ethyl 2-methylbutyrate is even more apple-like than ethyl 3-methylbutyrate, and is the dominant role in the aroma of ripe delicious apples.

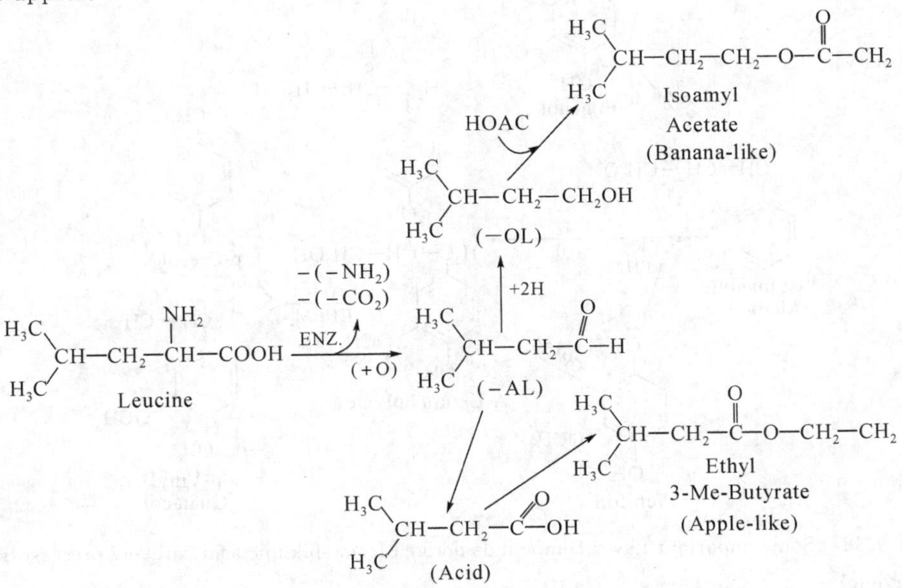

Fig. 9.10 Enzymatic conversion of leucine to volatiles illustrating the aroma compounds formed from amino acids in ripening fruits

9.3.7 Flavors Derived from the Shikimic Acid Pathway

In biosynthetic systems, the shikimic acid pathway provides the aromatic portion of compounds related to shikimic acid, and the pathway is best known in its role in the production of phenylalanine and the aromatic amino acids. In addition to flavor compounds derived from aromatic amino acids, the shikimic acid pathway provides other volatile compounds that are frequently associated with essential oils (Fig. 9.11). It also provides the phenyl propanoid skeleton to lignin polymers that are the main structural elements of plants. As indicated in Fig. 9.11, lignin yields many phenols during the pyrolysis, and the characteristic aroma of smokes used in foods is largely caused by compounds developed from precursors in the shikimic acid pathway.

Also apparent from Fig. 9.11 is that vanillin, the most important characterizing compound in vanilla extracts, can be obtained via this pathway or as a lignin by-product during the processing of wood pulp and paper. Vanillin is also biochemically synthesized in the vanilla bean, where it initially is present largely as vanillin glucoside until the glycoside is hydrolyzed during the fermentation. The methoxylated aromatic rings of the pungent principles in ginger, pepper, and chili peppers, discussed earlier in this chapter, also contain the essential features of those compounds in Fig. 9.11. Cinnamyl alcohol is an aroma constitutent of cinnamon spice, and eugenol is the principal aroma and pungency element in cloves.

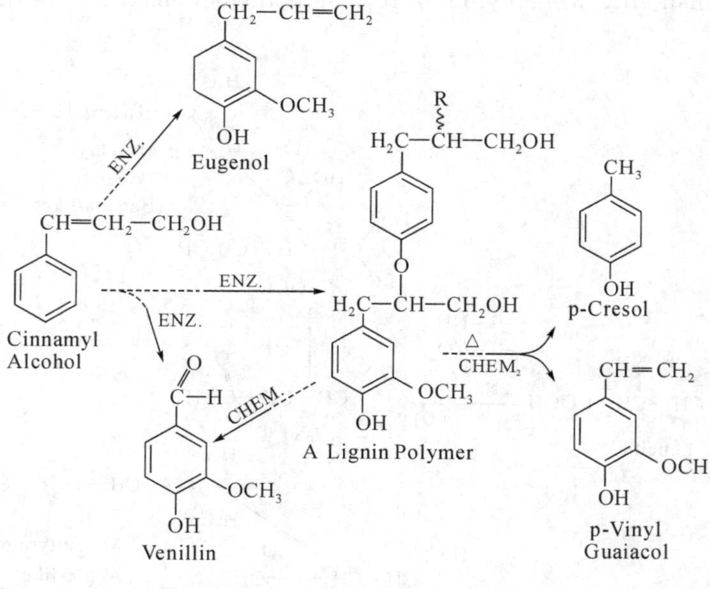

Fig. 9.11 Some important flavor compounds derived from shikimic acid pathway precursors (chem. =nonenzymic)

9.3.8 Volatile Terpenoids in Flavors

Because of the abundance of terpenes in plant materials used in the essential oil and perfumery industries, their importance in other plant-associated flavors is sometimes underestimated. They are largely responsible, however, for the flavors of citrus fruits and many seasonings and herbs. Terpenes are present in low concentrations in several fruits, and are responsible for much of the flavor of raw carrot roots. Terpenes are biosynthesized through the isoprenoid (C_5) paths (Fig. 9.12), and monoterpenes contain 10 C atoms; the sesquiterpenes contain 15 C atoms.

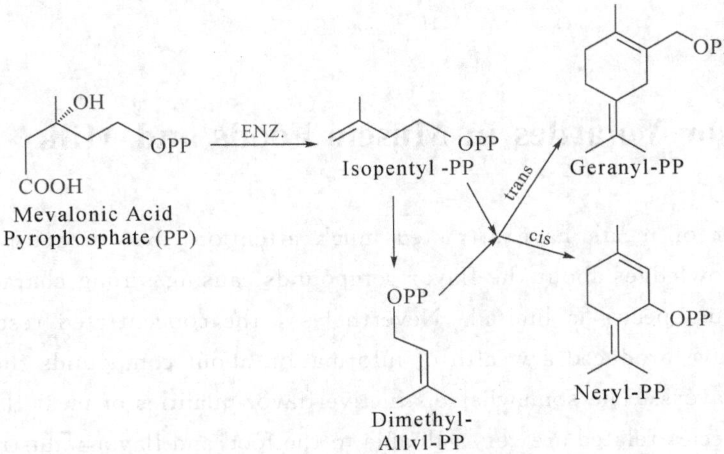

Fig. 9.12 Generalized isoprenoid scheme for the biosynthesis of monoterpenes

Sesquiterpenes are also important characterizing aroma compounds, and b-sinensal (XIX) and nootkatone (XX) serve as good examples because they provide characterizing flavors to oranges and grapefruit, respectively. The diterpenes (C_{20}) are too large and nonvolatile to contribute directly to aromas.

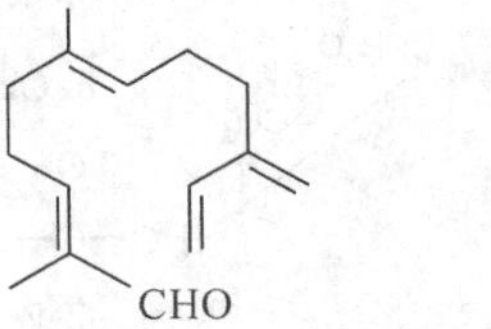

{XIX} β-Sinesal

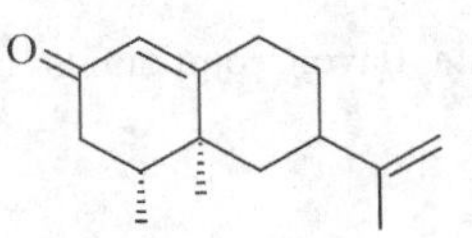

(XX) Nootkatone

Terpenes frequently possess extremely strong character-impact properties, and many can be easily identified by one experienced with natural product aromas. Optical isomers of terpenes, as well as optical isomers of other nonterpenoid compounds, can exhibit extremely different odor qualities. In the case of terpenes, the carvones have been studied extensively from this perspective, and the aroma of d-carvone [4S-(+)carvone] (XXI) has the characteristic aroma of caraway spice; l-carvone [4R-(-)carvone] (XXII) possesses a strong, characteristic spearmint aroma. Studies on such pairs of compounds are of interest since they provide information on the fundamental process of olfaction and structure-activity relationships for molecules.

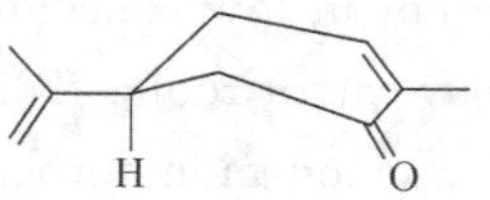

{XXI} 4S-(+)Carvone

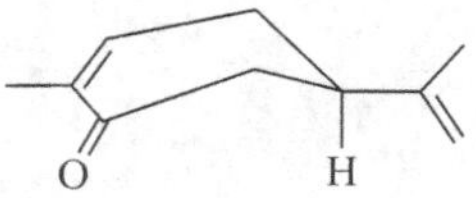

(XXII) 4R-(−)Carvone

9.4 Flavor Volatiles in Muscle Foods and Milk

The flavor of meats have attracted much attention, but in spite of considerable researches, knowledges about the flavor compounds causing strong character impacts for meats of various species is limited. Nevertheless, the concentrated research efforts on meat flavors have produced a wealth of information about compounds that contribute to cooked meat flavors. The somewhat distinctive flavor qualities of meat flavor compounds that are not species-related are very valuable to the food and flavor industry, but chemical functions of lightly cooked and species-related flavors are still eagerly sought.

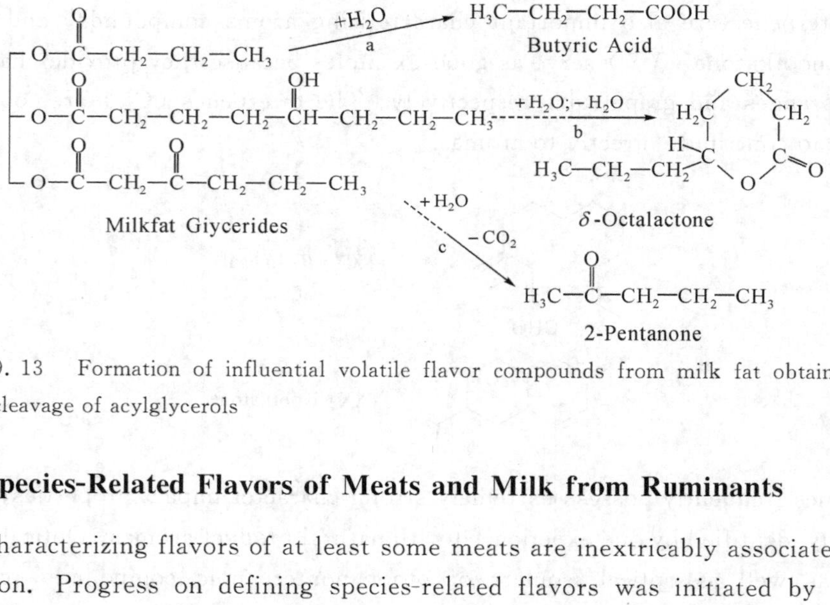

Fig. 9.13 Formation of influential volatile flavor compounds from milk fat obtained through hydrolytic cleavage of acylglycerols

9.4.1 Species-Related Flavors of Meats and Milk from Ruminants

The characterizing flavors of at least some meats are inextricably associated with the lipid fraction. Progress on defining species-related flavors was initiated by Wong and coworkers in relation to lamb and mutton flavors. These workers showed that a characteristic sweat-like flavor of mutton was closely associated with some volatile, medium chain-length fatty acids of which some methyl-branched members are highly significant. The formation of one of the most important branched-chain fatty acids in lamb and mutton, 4-methyloctanoic acid, is shown in Fig. 9.14. Ruminal fermentations yield acetate, propionate, and butyrate, but most fatty acids are biosynthesized from acetate, which yields nonbranched chains. Some methyl branching occurs routinely because of the presence of propionate, but when dietary and other factors enhance the propionate concentrations in the rumen, greater methyl branching occurs. Ha and Lindsay have shown that several medium-chain, methyl-branched fatty acids are important contributors to species-related flavors, and 4-ethyloctanoic acid (threshold = 1.8 ppb in water) is particularly important for conveying goat-like flavors to both meat and milk products. Additionally, several alkylphenols (methylphenol isomers, isopropylphenol isomers, and methyl-isopropylphenol isomers) contribute very characteristic cow-like and sheep-like species-related flavors to meats and milk. Alkylphenols are present as free and conjugate-bound substances in meats and milk, and are derived from shikimic acid pathway biochemical intermediates found in forages. Sulfate (XXV), phosphate (XXVI), and glucuronide (XXVII) conjugates of alkylphenols are formed in vivo, and are distributed via the circulatory system. Both enzymic and thermal hydrolysis of conjugates release phenols that enhance flavor development during the fermentation and cooking of meat and dairy products.

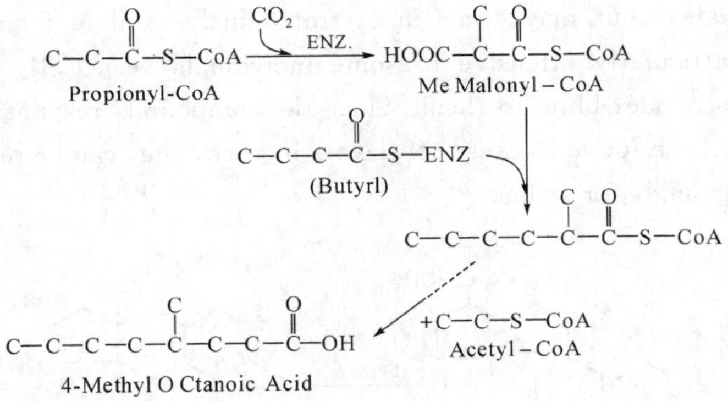

Fig. 9.14 Ruminant biosynthesis of methyl-branched, medium-chain fatty acids, ldehydes and acids

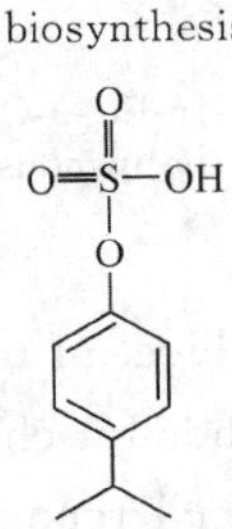

(XXV) p-Isopropylphenyl Sulfate

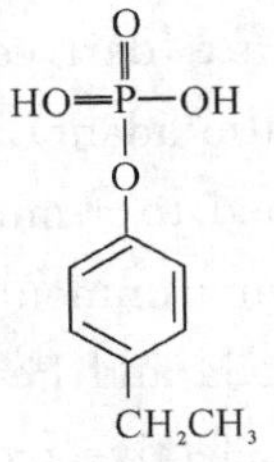

(XXVI) p-Ethylphenyl Phosphate

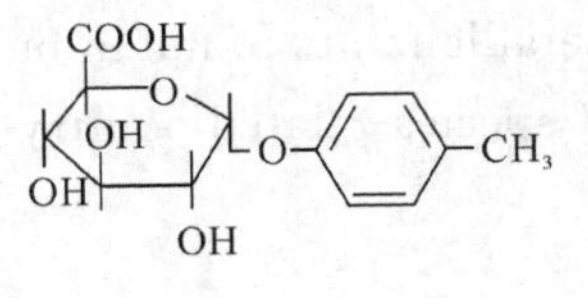

(XXVII) p-Crosyl Glucuronide

9.4.2 Species-Related Flavors of Meats from Nonruminants

Species-related aspects of the flavor chemistry of nonruminant meats remain somewhat incomplete. However, studies have shown that the g-C_5, C_9, and C_{12} lactones are reasonably abundant in pork, and these compounds likely contribute to some of the sweet-like flavor of pork. The distinct pork-like or piggy flavor, noticeable in lard or cracklings and in some pork, is caused by p-cresol and isovaleric acid that are produced from microbial conversions of corresponding amino acids in the lower gut of swine. Similar formation of indole and skatole from tryptophan may also intensify unpleasant piggy flavors in pork.

Much interest has centered on the aroma compounds responsible for the swine sex odor that causes serious off flavors in pork. Two compounds are responsible for the flavor, 5a-androst-16-en-3-one (Fig. 9.15), which has a urinous aroma, and 5a-androst-16-en-3a-ol, which has a musklike aroma. The swine sex odor compounds are mainly associated with males, but may occur in castrated males and in females. The steroid compounds are particularly offensive to some individuals, especially women, and yet others are genetically odor-blind to them. Since the compounds responsible for the swine sex odor have only been found to cause off flavors in pork, they can be regarded as species-related flavor compounds for swine.

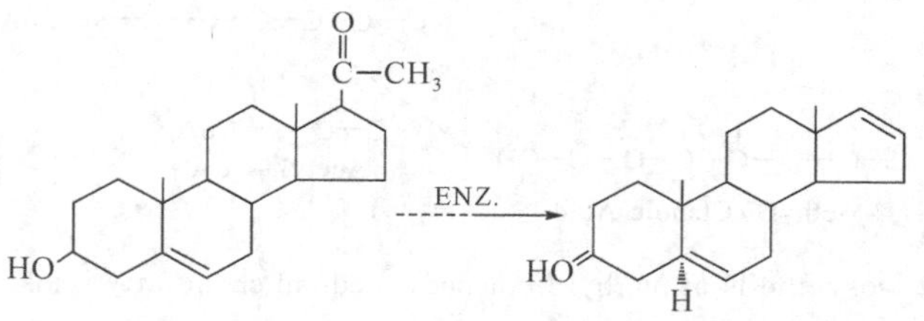

Fig. 9.15 Formation of the steroid compound responsible for the urinous aroma associated with the swine sex odor defect of pork

The distinctive flavors of poultry have also been the subject of many studies, and the lipid oxidation appears to yield the character impact compounds for chicken. The carbonyls c-4-decenal, t-2-c-5-undecadienal, and t-2-c-4-t-7-tridecatrienal reportedly may contribute the characteristic flavor of stewed chicken, and they are derived from lineolic and arachidonic acids. Chickens accumulate a-tocopherol (an antioxidant), but turkeys do not, and during cooking, such as roasting, carbonyls are formed to a much greater extent in turkey than in chicken. Additionally, certain chemical environments greatly affect the outcome of some lipid autoxidations. For example, Swoboda and Peers have shown that the presence of copper ions and a-tocopherol results in selective oxidations of milkfat, producing octa-1,c-5-dien-3-one, the cause of metallic faints found in butter. Directed lipid oxidations may also occur in the development of species-related poultry flavors, leading to species-related flavors.

9.4.3 Volatiles in Fish and Seafood Flavors

Characterizing flavors in seafoods cover a somewhat broader range of flavor qualities than those occurring in other muscle foods. The broad range of animals involved (finfish, shellfish, and crustaceans) and the variable flavor and aroma qualities related to freshness each account for the different flavors encountered. Historically, trimethylamine has been associated with fish and crab-like aromas, and alone it exhibits an ammoniacal, fishy aroma. Trimethylamine and dimethylamine are produced through the enzymic degradation of trimethylamine oxide (Fig. 9.16), which is found in significant quantities only in saltwater species of seafoods. Since very fresh fish contain essentially no trimethylamine, this compound modifies and contributes to the aroma of staling fish, in which it enhances fishhouse-type aromas. Trimethylamine oxide serves as a part of the buffer system in marine fish species. The formaldehyde produced concurrently with dimethylamine is believed to facilitate protein cross-linking, and thereby contributes to the toughening of fish muscle during the frozen storage.

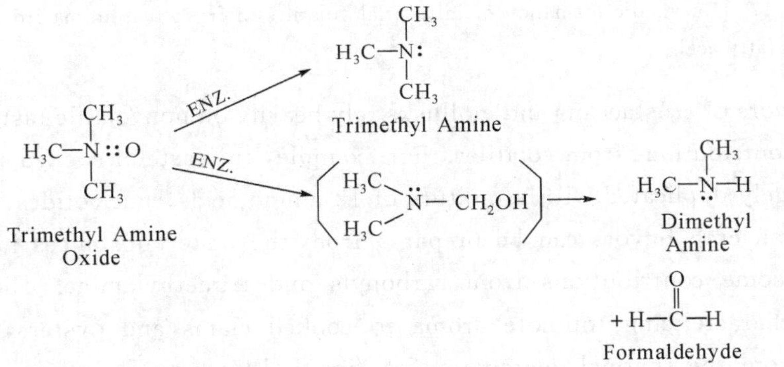

Fig. 9.16 Microbial formation of principal volatile amines in fresh saltwater fish species

Fishy aromas characterized by such terms as "oxidized fish oil" and "cod liver oil-like" are largely caused by carbonyl compounds produced from the autoxidation of long-chain-3-unsaturated fatty acids. These characteristic aromas result from 2, 4, 7-decatrienal isomers, and c-4-heptenal potentiates the fishy character of the decatrienals.

Because the fresh flavors and aromas of seafoods frequently have been greatly diminished or lost from fresh, frozen, and processed products available through commercial channels, many consumers associate the fishy flavors just described with all fish and seafoods. However, very fresh seafoods exhibit delicate aromas and flavors quite different from those usually evident in "commercially fresh" seafoods. It has been discovered that a group of enzymatically derived aldehydes, ketones, and alcohols provides the characterizing aroma of fresh fish, and these are very similar to the C_6, C_8, and C_9 compounds produced by plant lipoxygenases (see Section 9.3.5.1). Collectively, these compounds provide melony, heavy plant-like, and fresh fish aromas, and they are derived

FOOD CHEMISTRY

through a lipoxygenase system. The lipoxygenase systems found in fish and seafoods perform enzymic oxidations related to leukotriene synthesis, and the flavor compound production is a by-product of those reactions.

Hydroperoxidation followed by disproportionation reactions apparently leads first to the alcohol (Fig. 9.17), and then to the corresponding carbonyl. Some of these contribute to the distinctive flavors of very fresh cooked fish, either directly or as reactants that lead to new flavors during cooking.

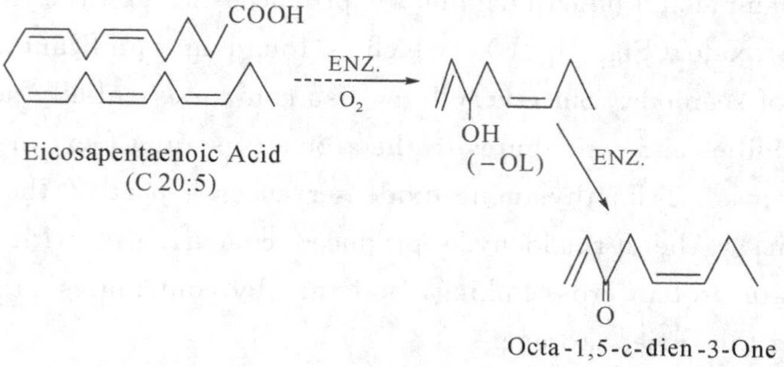

Fig. 9.17 Enzymatic formation of influential volatiles in fresh fish aroma from long-chain-3-unsaturated fatty acid

The flavors of crustaceans and mollusks rely heavily on nonvolatile taste substances in addition to contributions from volatiles. For example, the taste of cooked snow crab meat has been largely duplicated with a mixture of 12 amino acids, nucleotides, and salt ions. Good imitation crab flavors can be prepared from the taste substances just mentioned, along with some contributions from carbonyls and trimethylamine. Dimethyl sulfide provides a characterizing top-note aroma to cooked clams and oysters, and it arises principally from the thermal degradation of dimethyl-b-propiothetin present in ingested marine microflora.

9.5 Development of Process or Reaction Flavor Volatiles

Many flavor compounds found in cooked or processed foods occur as the result of reactions common to all types of foods, regardless of whether they are of animal, plant, or microbial derivation. These reactions take place when suitable reactants are present and appropriate conditions (heat, pH, and light) exist. Process or reaction flavors are discussed separately in this section because of their broad importance to all foods, and because they comprise a large volume of natural flavor concentrates that are used widely in foods, especially when meat or savory flavors are desired. Related information can be found in discussions dealing with carbohydrates, lipids, and vitamins.

9.5.1 Thermally Induced Process Flavors

Traditionally, these flavors have been broadly viewed as products from browning reactions because of early discoveries showing the role of reducing sugars and amino acid compounds in the induction of a process that ultimately leads to the formation of brown pigments. Although browning reactions are almost always involved in the development of process flavors in foods, the interactions between (a) degradation products of the browning reaction and (b) other food constituents are also important and extensive. By taking a broad approach to discussions of thermally induced flavors, the aforementioned interactions, as well as reactions that occur following the heat treatment, can be appropriately considered.

Although many of the compounds associated with process flavors possess potent and pleasant aromas, relatively few of these compounds seem to provide truly distinguishing character-impact flavor effects. Instead they often contribute general nutty, meaty, roasted, toasted, burnt, floral, plant, or caramel odors. Some process flavor compounds are acyclic, but many are heterocyclic, with nitrogen, sulfur, or oxygen substituents common (Fig. 9.18). These process flavor compounds occur in many foods and beverages, such as roasted meats, boiled meats, coffee, roasted nuts, beer, bread, crackers, snack foods, cocoa, and most other processed foods. The distribution of individual compounds does, however, depend on factors such as the availability of precursors, temperature, time, and water activity.

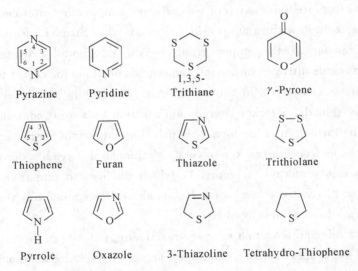

Fig. 9.18 Some heterocyclic skeletons found commonly in flavor compounds associated with thermally-induced or browning flavors

Production of process flavor concentrates is accomplished by selecting reaction mixtures and conditions so that those reactions occurring in the normal food processing are duplicated. Selected ingredients (Table 9.1), usually including a reducing sugar, amino acids, and compounds with sulfur atoms, are processed under elevated temperatures to produce a distinctive profile of flavor compounds. Thiamin is a popular ingredient because it provides both nitrogen and sulfur atoms already in ring structures.

Table 9.1 Some common ingredients used in process flavor reaction systems of the development of meat-like flavors

Hydrolyzed vegetable protein	Thiamin
Yeast atuolysak	Cysteine
Beef extract	Glutathione
Specific animalfats	Glucose
Chicken egg solids	Arabinose
Glycerol	5'-Ribonucleotides
Monosodium glutamate	Methionine

Because of the large number of process flavor compounds produced during the normal food processing or process simulation, it is unrealistic to cover the chemistry of their formation in depth. Rather, examples are given to illustrate some of the more important flavor volatiles formed and the mechanisms of their formation. Alkyl pyrazines were among the first compounds to be recognized as important contributors to the flavors of all roasted, toasted, or similarly thermally processed foods. The most direct route of their formation results from the interaction of a-dicarbonyl compounds (intermediate products in the Maillard reaction) with amino acids through the Stecker degradation reaction (Fig. 9.19). Transfer of the amino group to the dicarbonyl provides a means for integrating amino acids nitrogen into small compounds destined for any of the condensation reaction mechanisms envisioned in these reactions. Methionine has been selected as the amino acid involved in the Strecker degradation reaction because it contains a sulfur atom and it leads to the formation of methional, which is an important characterizing compound in boiling potatoes and cheese-cracker flavors. Methional also readily decomposes further to yield methanethiol (methyl mercaptan), which oxidizes to dimethyl disulfide, thus providing a source of reactive, low-molecular-weight sulfur compounds that contribute to the overall system of flavor developments.

Hydrogen sulfide and ammonia are very reactive ingredients in mixtures intended for the development of process flavors, and they are often included in model systems and assist in determining reaction mechanisms. Thermal degradation of cysteine (Fig. 9.20) yields both ammonia and hydrogen sulfide as well as acetaldehyde. Subsequent reaction of acetaldehyde with a mercapto derivative of acetoin (from the Maillard reaction) gives rise to thiazoline, which contributes to the flavor of boiled beef.

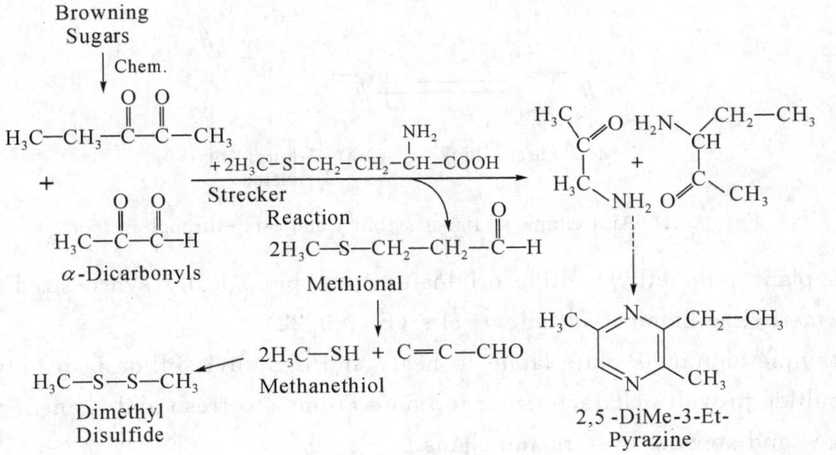

Fig. 9.19 Formation of an alkyl pyrazine and small sulfur compounds through reactions occurring in the development of process flavors

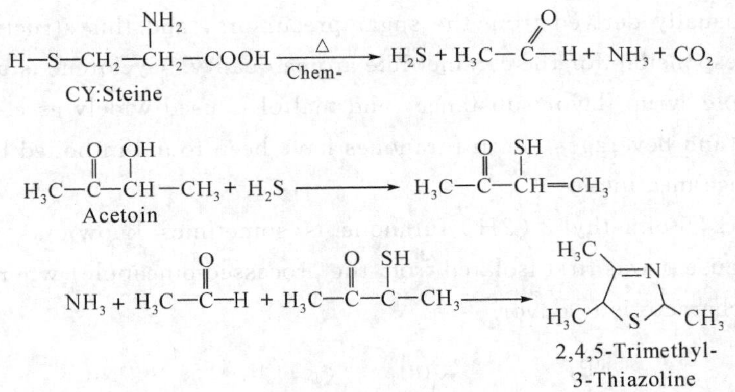

Fig. 9.20 Formation of a thiazoline found in cooked beef through the reaction of fragments from cysteine and sugar-amino browning (chem. = nonenzymic)

Some heterocyclic flavor compounds are quite reactive and tend to degrade or interact further with components of foods or reaction mixtures. An interesting example of flavor stability and carry-through in foods is provided by the compounds shown in Fig. 9.21, both of which provide distinct but different meat-like aromas. A roasting meat aroma is exhibited by 2-methyl-3-furanthiol (reduced form), but upon oxidation to the disulfide form, the flavor becomes more characteristics of fully cooked meat that has been held for some time. Chemical reactions, such as the one just mentioned, are responsible for the subtle changes in meat flavor that occur because of the degree of cooking and the time interval after cooking.

During processing of complex systems, sulfur as such, or as thiols or polysulfides, can be incorporated in various compounds, resulting in the generation of new flavors. However, even though dimethyl sulfide is often found in processed foods, it does not react

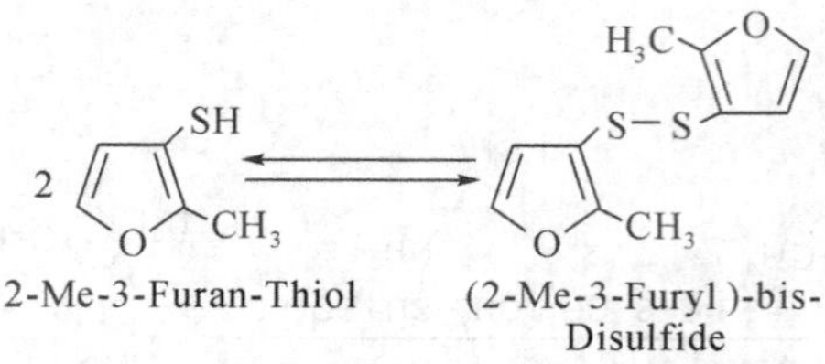

FOOD CHEMISTRY

Fig. 9.21 An example of flavor stability and carry-through in food

readily. In plants, dimethyl sulfide originates from biologically synthesized molecules, especiall S-methylmethionine sulfonium salts (Fig. 9.22).

S-Methylmethionine is quite labile to heat, and dimethyl sulfide is readily released. Dimethyl sulfide provides characterizing top-note aromas to fresh and canned sweet corn, tomato juice, and stewing oysters and clams.

Some of the most pleasant aromas derived from process reactions are provided by the compounds shown in Fig. 9.23. These compounds exhibit caramel-like aromas and have been found in many processed foods. The planar enol-cyclic-ketone structure (see also Fig. 9.21) is usually derived from the sugar precursors, and this structural component appears to be responsible for the caramel-like aroma quality. Cyclotene is used widely as a synthesized maple syrup flavor substance, and maltol is used widely as a flavor enhancer for sweet foods and beverages. Both furanones have been found in boiled beef where they appear to enhance meatiness.

4-Hydroxy-2,5-dimethyl-3(2H)-furanone is sometimes known as the "pineapple compound" because it was first isolated from the processed pineapple, where it contributes strongly to its characteristic flavor.

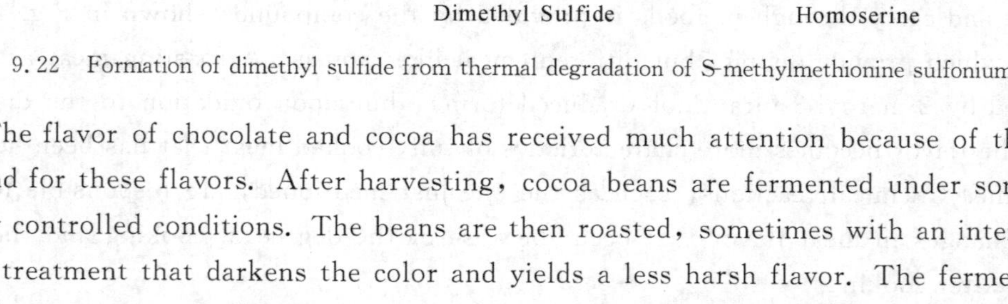

Fig. 9.22 Formation of dimethyl sulfide from thermal degradation of S-methylmethionine sulfonium salts

The flavor of chocolate and cocoa has received much attention because of the high demand for these flavors. After harvesting, cocoa beans are fermented under somewhat poorly controlled conditions. The beans are then roasted, sometimes with an intervening alkali treatment that darkens the color and yields a less harsh flavor. The fermentation hydrolyzes sucrose to reducing sugars, frees amino acids, and oxidizes some polyphenols. During roasting, many pyrazines and other heterocyclics are formed, but the unique flavor of cocoa is derived from an interaction between aldehydes from the Strecker degradation reaction. The reaction shown in Fig. 9.23 between phenylacetaldehye (from phenylalanine) and 3-methylbutanal (from leucine) constitutes an important flavor-forming reaction in cocoa. The product of this aldol condensation, 5-methyl-2-phenyl-2-hexenal, exhibits a characterizing persistent chocolate aroma. This example also serves to show that reactions in the development of process flavors do not always yield heterocyclic aroma compounds.

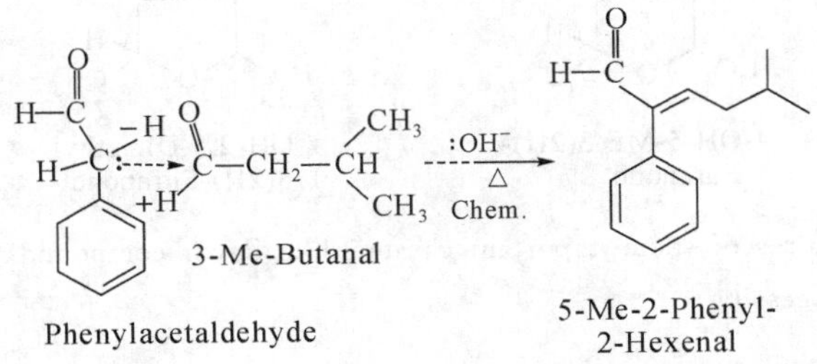

Fig. 9.23 Formation of an important cocoa aroma volatile through an aldol condensation of two Strecker reaction-derived aldehydes (chem. = nonenzymic)

9.5.2 Volatiles Derived from Oxidative Cleavage of Carotenoids

Oxidations focusing on triacylglycerols and fatty acids are discussed in another section, but some extremely important flavor compounds that are oxidatively derived from carotenoid precursors have not been covered and deserve attention here. Some of these reactions require the singlet oxygen through the chlorophyll sensitization; others are photooxidation processes. A large number of flavor compounds, derived from oxidizing carotenoids (or isoprenoids), have been identified in curing tobacco, and many of these are considered important for characterizing tobacco flavors. However, relatively few compounds in this category (three representative compounds are shown in Fig. 9.24) are currently considered highly important as food flavors. Each of these compounds exhibits unique sweet, floral, and fruit-like characteristics that vary greatly with concentration. They also blend nicely with aromas of foods to produce subtle effects that may be highly desirable or very undesirable. β-Damascenone exerts very positive effects on the flavors of wines, but in beer this compound at only a few parts per billion results in a stale, raisin-like note. β-Ionone also exhibits a pleasant violet, floral aroma compatible with fruit-type flavors, but it is also the principal off-flavor compound present in oxidized, freeze-dried carrots. Furthermore, these compounds have been found in black tea, where they make positive contributions to the flavor. Theaspirane and related derivatives contribute importantly to the sweet, fruity, and earthy notes of tea aroma. Although usually present in low concentrations, these compounds and related ones appear to be widely distributed, and it is likely that they contribute to the full, well-blended flavors of many foods.

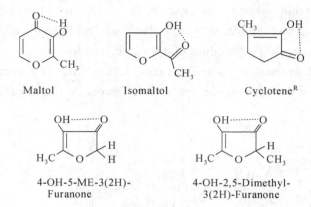

Fig. 9.24 Structures of some important caramel-like flavor compounds derived from reactions occurring during the processing

9.6 Future Directions of Flavor Chemistry and Technology

Knowledge about the chemistry and technology of flavors has expanded greatly in the past 35 years, and information has accumulated at this point where control and manipulation of many flavors in foods is possible. However, some of the areas of research are likely to be fruitful in the foreseeable future are binding of flavors to macromolecules, structure-activity relationships in taste and olfaction as determined using computer techniques, control of glycoside precursors of plant-derived flavors, control of reaction flavor chemistry, and flavor development as related to genetics of plant cultivars and cultured cells (microbial and tissue cultures).

The issue of authenticity of natural flavors continues to attract considerable attention, and research on the various aspects of analysis and structural relationships of closely related molecules can be expected to enhance knowledge about subtle molecular influences on the flavor quality of such optical isomers. Isotopic mass spectrometry (carbon-13 and deuterium) techniques have made it possible in many cases to differentiate between natural and synthetic molecules, but carbon-13 enrichment alterations of synthetic molecules can render such approaches for detecting adulteration invalid. However, the site-specific natural isotope fractionation measured by the nuclear magnetic resonance is developing to the point where it provides an isotropic fingerprint that makes it possible to detect the adulteration of natural origin flavors.

The chemical accuracy of nature-identical synthetic substances continues to be challenged by the new information that is accumulating on the unique odor and flavor properties of enantiomers and other chiral compounds. Such differences in the odor quality between enantiomers not only have relevance to nature-identical issues, but also impact strongly on basic understandings of structure-function relationships in the olfaction of molecules, and new information should further this field.

Enzymic production of flavors within foods and ingredients will undoubtedly form a significant area of flavor technology in the future. Early in the modern era of analytical flavor chemistry, that is, 35 years ago, the flavorese concept was offered as a means for regenerating flavors in processed foods, and this type of approach is certain to become more important. However, because of the complex nature of natural flavors as they are now understood, it becomes apparent that the flavorese concept as initially developed would meet with great difficulties when applied to foods. Still, with recent developments in making encapsulated flavor enzyme systems, it is possible to maintain substrate-enzyme proximity and to control the amounts of flavor compounds produced so that unbalanced flavors can be avoided. With the increasing emphasis on high-quality formulated, complex foods, flavor development by enzymes should find a role of increasing importance. Finally, efforts to identify flavor compounds will continue, particularly in cases in which important characterizing flavor compounds appear to be present.

Glossary

allium	葱属植物,大蒜
astringency	涩味
bitter	苦味
brussel sprouts	甘蓝小包菜
cruciferae	十字花科
eugenol	丁子香酚
flavors	风味物质
methionine	蛋氨酸
phenylacetaldehyde	苯乙醛
pungency	辛辣味
ruminants	反刍动物
salty	咸味
shiitake	香菇
sour	酸味
sweet	甜味
volatiles	挥发性物质

Chapter 10　Food Additive

10.1　Introduction

10.1.1　Definition

Food additives are any chemically synthesized or natural substances added to foods in order to improve the quality, color, flavor, or for the preserving and technological purpose. The use of food additives to conceal the damage or spoilage to foods or to deceive consumers is expressly prohibited by regulations governing the use of these substances in foods. In addition, food additive usages are discouraged where similar effects can be obtained by economical, good manufacturing practices.

10.1.2　Classification

A food additive is a substance, or a mixture of substances, added into food and involved in its production, processing, packaging and/or storage without being a major ingredient. Most of the additives or their degradation products generally remain in food, but in some cases they can be removed during the processing. The following examples illustrate and support the utilization of additives to enhance the:

10.1.2.1　Nutritive Values of Food

Additives such as vitamins, minerals, amino acids and amino acid derivatives are used to improve the nutritive values of food. A particular diet may also require the utilization of thickening agents, emulsifiers, sweeteners and so on.

10.1.2.2　Sensory Values of Food

Color, odor, taste and texture, which are of importance for the sensory value of food, may decrease during processing and storage. Such decreases can be corrected or readjusted by the addition of additives such as pigments, aroma compounds or flavor enhancers. Development of "off-flavor", for example, derived from the fat or oil oxidation, can be inhibited by antioxidants. Food texture can be stabilized by adding minerals or polysaccharides, and by many other means.

10.1.2.3 Shelf Life of Food

Longer shelf life has been demanded because of the current forms of food production and distribution, and the world food supply situation requires preservation by avoiding deterioration as much as possible. The extension of shelf life involves protection against microbial spoilage, for instance, by using antimicrobial additives and by using active agents which suppress and retard undesired chemical and physical changes in food. The latter is achieved by stabilization of pH using buffering additives or stabilization of texture with thickening or gelling agents, which are polysaccharides in most cases.

10.1.2.4 Practical Value

It is very necessary to increase the use of additives due to the common trend towards foods which are easy and quick to prepare (e.g. convenient foods).

10.2 Acids

Both organic and inorganic acids occur extensively in nature, where they function in a variety of roles ranging from intermediary metabolites to components of buffer systems. Acids are used for numerous purposes in foods and food processing, where they provide the benefits of many natural actions. One of the most important functions of acids in foods is participation in buffering systems. Beside this, acids are also used to regulate the pH of chemical leavening systems, to play the role of specific acidic microbial inhibitors (e.g. sorbic acid, benzoic acid) in the food preservation, and function as chelating agents. Acids are important in the setting of pectin gels, they serve as defoaming agents and emulsifiers, and they induce coagulation of milk proteins in the production of cheese and cultured dairy products such as sour cream. In natural culturing processes, lactic acid ($CH_3-CHOH-COOH$) produced by *streptococci* and *lactobacilli* causes coagulation by lowering the pH to near the isoelectric point of casein. Cheeses can be produced by the addition of rennet and acidulants, such as citric acid and hydrochloric acids, to cold milk (4-8 ℃). Subsequent warming of the milk (to 35 ℃) results in a uniform gel structure. Addition of acid to warm milk produces a protein precipitate rather than a gel.

δ-Gluconolactone can be used for slowing the acid production in cultured dairy products and chemical leavening systems, as it forms gluconic acid by slowly hydrolyzing in aqueous systems (Fig. 10.1). Dehydration of lactic acid yields lactide, a cylic dilactone (Fig. 10.2), which can be also used as a slow-release acid in aqueous systems. The dehydration reaction happens under conditions of low water activity and elevated temperature. Introduction of lactide into foods with high water activity causes a reversal of the process with the production of two moles of lactic acid.

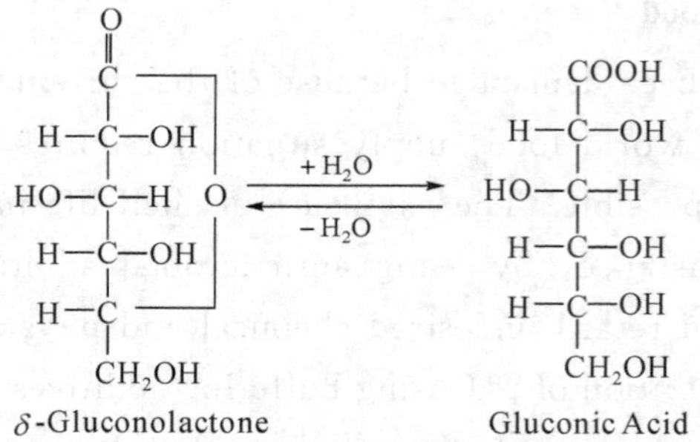

FOOD CHEMISTRY

Fig. 10.1 Formation of gluconic acid from the hydrolysis of δ-gluconolactone

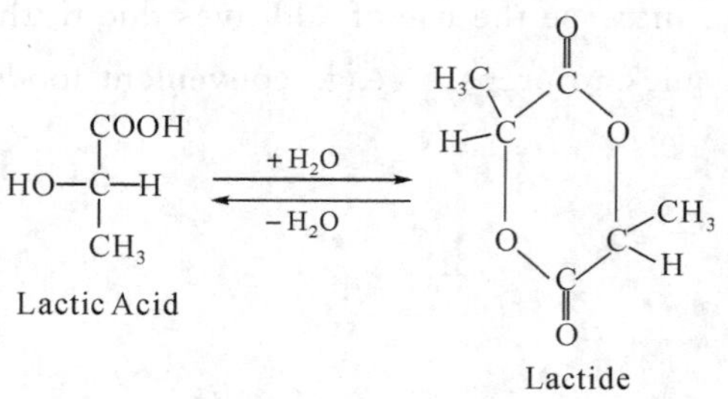

Fig. 10.2 Equilibrium reaction showing the formation of lactic acid from the hydrolysis of lactide

Acids such as citric are added to some moderately acidic fruits and vegetables to lower the pH to lower than 4.5. In canned foods, this permits the sterilization to be achieved under less severe thermal conditions than what is necessary for less acid products, and has the added advantage of precluding the growth of hazardous microorganisms *Clostridium botulinum*.

Acids, such as potassium acid tartrate, are utilized in the manufacture of fondant and fudge to induce limited hydrolysis or inversion of sucrose. Inversion of sucrose yields fructose and glucose, which improve texture through the inhibition of excessive growth of sucrose crystals. Monosaccharides can prevent crystallization by contributing to the complexity of the syrup and by lowering its equilibrium relative humidity.

One of the most important contributions of acids to foods is that they are able to form a sour or tart taste. Acids also have the ability to modify and intensify the taste perception of other flavoring agents. The hydrogen ion or hydronium ion (H_3O^+) is involved in the generation of the sour taste response. In addition, short-chain free fatty acids (C_2-C_{12}) contribute significantly to the aroma of foods. For example of butyric acid, at relatively high concentrations it improves strongly to the characteristic flavor of hydrolytic rancidity, while at lower concentrations it contributes to the typical flavor of products such as cheese and butter.

Numerous organic acids can be used in food industry. Some of the more commonly used acids are acetic (CH_3COOH), lactic (CH_3—$CHOH$—$COOH$), citric ($HOOC$—CH_2—$COH(COOH)$—CH_2—$COOH$), malic ($HOOC$—$CHOH$—CH_2—$COOH$), fumaric (HOOC—CH=CH—COOH), succinic (HOOC—CH_2—CH_2—COOH), and tartaric (HOOC—CHOH—CHOH—COOH). Phosphoric acid (H_3PO_4) is the only inorganic acid extensively employed as a food acidulant. Phosphoric acid is an important acidulant in flavored carbonated beverages, particularly in colas. The other inorganic acids such as HCL and H_2SO_4 are usually too highly dissociated for food applications, and their use may lead to problems with quality attributes of foods. Dissociation constants for some acids used in food are shown in Table 10.1.

Table 10.1 Dissociation constants at 25 ℃ for some acids used in foods

Acid	Step	pK_a	Acid	Step	pK_a
Organic acids					
Acetic		4.75	Propionic		4.87
Adipic	1	4.43	Succinic	1	4.16
	2	5.41		2	5.61
Benzoic		4.19	Tartaric	1	3.22
n-Butyric		4.81		2	4.82
Citric	1	3.14	Inorganic acids		
	2	4.77	Carbonic	1	6.37
	3	6.39		2	10.25
Formic		3.75	o-Phosphoric	1	2.12
Fumaric	1	3.03		2	7.21
	2	4.44		3	12.67
Hexanoic		4.88	Sulfuric	2	1.92
Lactic		3.08			
Malic	1	3.40			
	2	5.10			

10.3 Bases

Basic or alkaline substances are used in various applications in foods and food processing. Although the majority of applications involve buffering and pH adjustments, other functions include carbon dioxide evolution, improvement of color and flavor, solubilization of proteins, and chemical peeling.

Alkali treatments are imposed on several food products for the purpose of color and flavor improvement. Solutions of sodium hydroxide (0.25%-2.0%) are added into ripe olives to aid in the removal of the bitter principal and to develop a darker color. Soy proteins are solubilized through alkali processings, and the concern has been expressed about alkaline-induced racemization of amino acids and losses of other nutrients. Small amounts of sodium bicarbonate are used in the manufacture of peanut brittle candy to enhance the caramelization and browning, and to provide, through release of carbon dioxide, a somewhat porous structure. Bases, usually potassium carbonate, are also used in the cocoa processing for the production of dark chocolate. The elevated pH enhances sugar-amino browning reactions and polymerization of flavonoids, leading to a smoother, less acid and less bitter chocolate flavor, a darker color, and a slightly improved solubility.

It is sometimes necessary for food systems to adjust higher pH values to achieve more stable or more desirable. For instance, alkaline salts such as disodium phosphate, trisodium phosphate and trisodium citrate are normally used in the preparation of processed cheese (1.5%-3%) to increase the pH (5.7-6.3) and to influence the protein dispersion. This salt-protein interaction improves the emulsifying and water-binding capabilities of the cheese proteins since the salts bind the calcium components of the casein micelles forming chelates.

The addition of phosphates and citrates changes the salt balance in fluid milk products by forming complexes with calcium and magnesium ions from casein. The mechanism is incompletely understood, but depending on the type and concentration of salt added, the milk protein system can undergo stabilization, gelation, or destabilization.

Strong bases are employed for peeling fruits and vegetables. Exposure of the products to hot solutions at 60-82 ℃ (140-180 ℉) of sodium hydroxide (approximately 3%), with subsequent mild abrasion, effects peel removal with substantial reductions in plant wastewater as compared with the conventional peeling techniques. Differential solubilization of cell and tissue constituents (pectic substances in the middle lamella are particularly soluble) provides the basis for caustic peeling processes.

10.4 Buffer Systems and Salts

10.4.1 Buffers and pH Control in Foods

Since most food are complex materials of biological origin, which include proteins, organic acids, and weak inorganic acid-phosphate salts, they contain many substances that can participate in pH control and buffering systems. Lactic acid and phosphate salts, together with proteins, are important for pH control in animal tissues; polycarboxylic acids, phosphate salts, and proteins are important in plant tissues. In plants, buffering systems containing citric acid (lemons, tomatoes, and rhubarb), malic acid (apples, tomatoes, and lettuce), oxalic acid (rhubarb, lettuce), and tartaric acid (grapes, pineapple) are common, and they usually function in conjunction with phosphate salts in maintaining pH control. Milk acts as a complex buffer because of this content of carbon dioxide, proteins, phosphate, citrate, and several other minor constituents.

When the pH must be altered, it is usually desirable to use a buffer system. This is accomplished naturally when lactic acid is produced in cheese and pickle fermentations. Also, in some cases where substantial amounts of acids are used in foods and beverages, it is desirable to reduce the sharpness of acid tastes, and obtain smoother product flavors without inducing neutralization flavors. This usually can be accomplished by establishing a buffer system in which the salt of a weak organic acid is predominant. The common ion effect is the basis for obtaining pH control in these systems, and the system develops as the added salts contain an ion that is already present in an existing weak acid. The added salt immediately ionizes causing repressed ionization of the acid with reduced acidity and a more stable pH. The efficiency of a buffer depends on the concentration of the buffering substances. Since there is a pool of undissociated acid and dissociated salt, buffers resists changes in pH. For instance, relatively large additions of a strong acid, such as hydrochloric acid, to an acetic sodium acetate system causes hydrogen ions to react with the acetate ion pool to increase the concentration of slightly ionized acetic acid, and the pH remains relatively stable. In a similar manner, addition of sodium hydroxide causes hydroxyl ions to react with hydrogen ions to form water molecules which were undissociated.

Titration of buffered systems and resulting titration curves (i.e., pH vs volume of base added) reveal their resistance to the pH change. If a weak acid buffer is titrated with a base, there is a gradual but steady increase in the pH as the system gets near to the neutralization; that is, the change in pH per milliliter of added base is small. Weak acids are only slightly dissociated at the beginning of the titration. However, the addition of hydroxyl ions shifts the equilibrium to the dissociated species and eventually the buffering capacity is overcome.

10.4.2 Salts in Processed Dairy Foods

Salts are widely used in processed cheeses and imitation cheeses to induce a uniform, smooth texture. These additives are sometimes referred to as emulsifying salts because of their ability to help dispersion of fat and oil. Although the emulsifying mechanism remains somewhat less than fully defined, anions from the salts when added to processed cheese combine with and remove calcium from the para-casein complex, and this causes rearrangement and exposure of both polar and nonpolar regions of the cheese proteins. It is also believed that the anions of these salts participate in ionic bridges between protein molecules, and provide a stabilized matrix that entraps the fat and oil in processed cheese. Salts used for cheese processings include mono-, di-, and tri-sodium phosphate, dipotassium phosphate, sodium acid pyrophosphate, sodium gexametaphosphate, tetrasodium pyrophosphate, and other condensed phosphates, trisodlum citrate, sodium tartrate, tripotassium citrate, and sodium potassium tartrate.

The certain phosphates such as trisodium phosphate added into milk, prevent separation of the milk fat and aqueous phases. The required amount varies with the season of the year and the source of milk. Concentrated milk sterilized by a high temperature, short-time method frequently gels upon storage. The addition of polyphosphates, such as sodium tripolyphosphate and sodium hexametaphosphate, prevents the gel formation through a protein denaturation and solubilizadon mechanism which involves complexing of calcium and magnesium by phosphates.

10.5 Antimicrobial Agents

10.5.1 Acid Antimicrobial

10.5.1.1 Sorbic Acid

Straight chain, monocarboxylic, and aliphatic fatty acids possess antimycotic activity, and α-unsaturated fatty acid analogs are especially effective. Sorbic acid (C—C=C—C=C—COOH) and its sodium and potassium salts are used extensively to inhibit mold and yeasts in a variety of foods including cheese, baked products, fruit juices, wine and pickles. Sorbic acid has a strong effect on preventing mold growth, and it contributes little flavor at the concentrations of up to 0.3 wt%. The method of application may involve direct incorporation, surface coatings, or incorporation in a wrapping material. The activity of sorbic acid increases as the pH decreases, indicating that the undissociated form is more inhibitory than the dissociated form. Generally, sorbic acid is effective up to pH 6.5, which is considerably above the effective pH ranges for propionic and benzoic acids.

The antimycotic action of sorbic acid appears to arise because molds are unable to metabolize the α-unsaturated diene system of its aliphatic chain. It has been proposed that the diene structure of sorbic acid interferes with cellular dehydrogenases that normally dehydrogenate fatty acids as the first step in the oxidation. Saturated short-chain (C_2-C_{12}) fatty acids are also moderately inhibitory to many molds, such as *Penicilliurn roqueforfi*. However, some of these molds are capable of mediating β-oxidation of saturated fatty acids to corresponding β-keto acids, especially when the concentration of the acid is only marginally inhibitory. Decarboxylation of the resulting β-keto acid yields the corresponding methyl ketone (Fig. 10.3) which does not possess antimicrobial properties. A few molds have also been shown to metabolize sorbic acid, and it has been suggested that this metabolism proceeds through β-oxidation, similar to that in mammals. All evidence indicates that animals and humans metabolize sorbic acid in nearly the same ways as they do in other naturally occurring fatty acids.

Although sorbic acid might at first appear quite stable and unreactive, it is quite often microbiologically or chemically altered in foods. Two other mechanisms for deactivating the antimicrobial properties of sorbic acid are shown in Fig. 10.4. The reaction labeled "a" has been demonstrated in molds, especially *P. roqueforti*. This involves direct decarboxylation of sorbic acid to yield the hydrocarbon 1,3-pentadierte, the intense aroma of which can cause gasoline or hydrocarbon-like off flavors when mold growth occurs in the presence of sorbic acid.

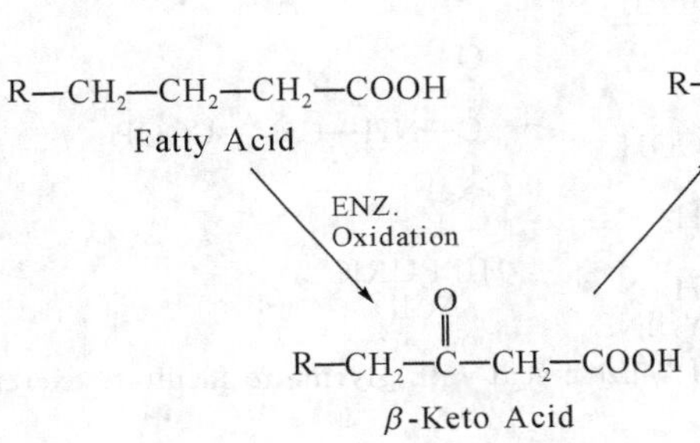

$$R-CH_2-CH_2-CH_2-COOH \quad\quad R-CH_2-\overset{O}{\underset{\|}{C}}-CH_3-CO_2$$
$$\text{Fatty Acid} \quad\quad\quad\quad \text{Methyl Ketone}$$

$$\xrightarrow{\text{ENZ. Oxidation}}$$

$$R-CH_2-\overset{O}{\underset{\|}{C}}-CH_2-COOH$$
$$\beta\text{-Keto Acid}$$

Fig. 10.3 Formation of a methyl ketone via mold-mediated enzymic oxidation of a fatty acid followed by a decarboxylation reaction

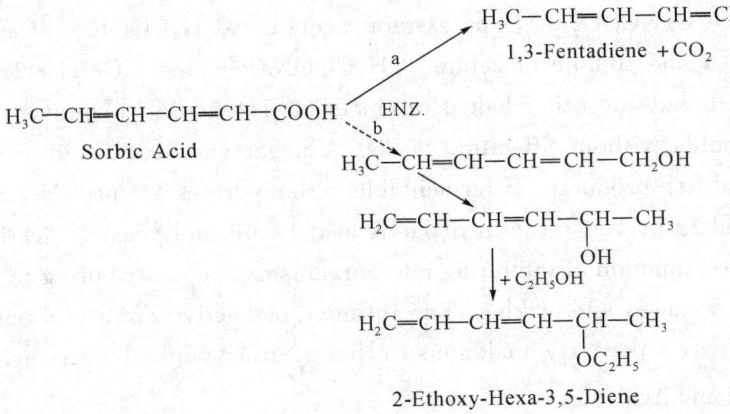

$$H_3C-CH=CH-CH=CH_2$$
$$\text{1,3-Fentadiene} + CO_2$$

$$H_3C-CH=CH-CH=CH-COOH \xrightarrow{\text{ENZ.}}$$
$$\text{Sorbic Acid}$$

$$H_3C-CH=CH-CH=CH-CH_2OH$$

$$H_2C=CH-CH=CH-\underset{OH}{\overset{}{CH}}-CH_3$$

$$\downarrow +C_2H_5OH$$

$$H_2C=CH-CH=CH-\underset{OC_2H_5}{\overset{}{CH}}-CH_3$$
$$\text{2-Ethoxy-Hexa-3,5-Diene}$$

Fig. 10.4 Enzymic conversions destroying the antimicrobial properties of sorbic acid: (a) decarboxylation carried out by *Penicillium sp.*; (b) formation of ethoxylated diene hydrocarbon in wine resulting from a reduction of the carboxyl group followed by the rearrangement and development of an ether

10.5.1.2 Benzoic Acid

Benzoic acid (C_6H_5COOH) has been widely used as an antimicrobial agent in food to inhibit the microorganisms, and this occurs naturally in cranberries, prunes, cinnamon, and cloves. The undissociated acid is the form with antimicrobial activity, and it exhibits the optimum activity in the pH range of 2.5-4.0, making it well suited for the use in acid foods such as fruit juices, carbonated beverages, pickles, and sauerkraut. The sodium salt of benzoic acid form is usually used, since it is more soluble in the water than the acid form. Once in the products, some of the salt converts to the active acid form, which is the most active against yeasts and bacteria and the least active against molds. Often benzoic acid is used with sorbic acid or parabens, and levels of use usually range from 0.05 to 0.1 wt%.

Benzoic acid has been found to cause no deleterious effects in humans when used in small amounts. It is readily eliminated from the body after conjugation with glycine (Fig. 10.5) to form hippuric acid (benzoyl glycine). This detoxification step precludes accumulation of benzoic acid in the body.

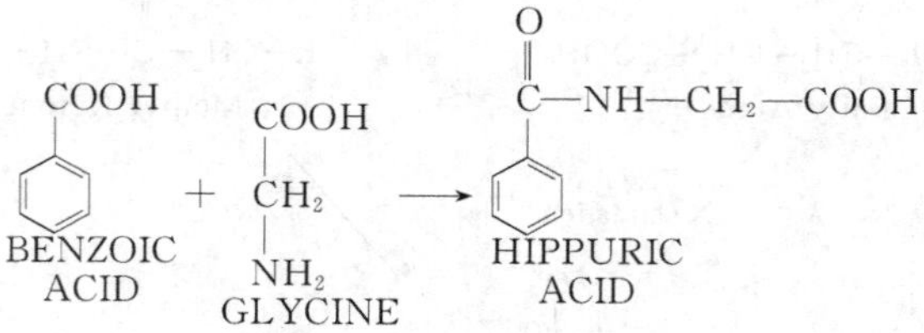

Fig. 10.5 Conjugation of benzoic acid with glycine to facilitate excretion

10.5.1.3 Acetic Acid

The preservation of foods with acetic acid (CH_3COOH) in the form of vinegar dates to antiquity. In addition to vinegar (4% acetic acid) and acetic acid, also used in food are sodium acetate (CH_3COONa), potassium acetate (CH_3COOK), calcium acetate [$(CH_3COO)_2Ca$], and sodium diacetate ($CH_3COONa \cdot CH_3-COOH \cdot 1/2H_2O$). The salts are used in bread and other baked products (0.1%-0.4%) to prevent ropiness and the growth of molds without affecting yeasts. Vinegar and acetic acid are employed in pickled meats and fish products. If fermentable carbohydrates are present, at least 3.6% acid must be used to prevent growth of lactic acid bacilli and yeasts. Acetic acid is also employed as a dual function of inhibiting microorganisms and contributing to flavor, in the food such as mayonnaise, and pickles. The antimicrobial activity of acetic acid increases as the pH is decreased, a property analogous to that of other aliphatic fatty acids.

10.5.1.4 Propionic Acid

Propionic acid (CH_3-CH_2-COOH) and its sodium and calcium salts have antimicrobial activity against molds and a few bacteria. This compound occurs naturally in Swiss cheese (up to 1 wt%), where it is produced by *Propionibacterium shermanii*. Propionic acid has found extensively use in the bakery field where it not only inhibits molds effectively, but also is active against the ropy bread organism, *Bacillus mesentericus*. Levels of use of propionic acid generally range up to 0.3% by weight. As with antimicrobial agents of other carboxylic acid, the undissociated form of propionic acid is active, and the range of effectiveness extends up to pH 5.0 in most applications. The toxicity of propionic acid to molds and certain bacteria is related to the inability of affected organisms to metabolize the three carbon skeleton. In mammals, propionic acid is metabolized in a manner similar to that of other fatty acids, and it has not been shown to cause any toxic effects at the levels utilized.

10.5.2 Ester Antimicrobial

10.5.2.1 Glyceryl Esters

Many fatty acids and monoglycerides show excellent antimicrobial activities against gram-positive bacteria and some yeasts. Unsaturated members, especially those with 18 carbon atoms, have strong activity as fatty acids; the medium chain-length members (12 carbon atoms) are most inhibitory when esterified to glycerol. Glyceryl monolaurate is inhibitory against several potentially pathogenic *Staphylococcus* and *Streptococcus* at concentrations of 15-250 ppm. It is commonly used in cosmetics, and can be used in some foods because of its lipid nature.

$$CH_2-O-\overset{O}{\underset{}{C}}-(CH_2)_{10}-CH_3$$
$$|$$
$$CHOH$$
$$|$$
$$CH_2OH$$

GLYCERYL MONOLAURATE

Lipophilic agents of this kind also exhibit inhibitory activity against *C. botulinum* in cured meats and in refrigerated, packaged fresh fish. The inhibitory effect of lipophilic glyceride derivatives apparently relates to their ability to facilitate the conduction of protons through the cell membranes, which effectively destroys the proton motive force that is necessary for the substrate transport. Cell-killing effects are observed only at high concentrations of these compounds, and death apparently results from the generation of holes in cell membranes.

10.5.2.2 Paraben Esters

The alkyl esters of p-hydroxybenzoic acid (PHB; parabens) are quite stable. Their solubility in water decreases with increasing alkyl chain length (methyl → butyl). The esters are mostly soluble in 5% NaOH.

The esters have strong effects against yeast, but less against bacteria, especially those gram-negative bacteria. The activity increases with increasing alkyl chain length. However, lower members of the homologous series are preferred because of better solubility.

Unlike benzoic acid, the paraben esters can be used over a wide pH range since their activity is almost independent of pH. As additives, they are usually applied at 0.3%-

0.06% as aqueous alkali solutions or as ethanol or propylene glycol solutions in fillings for baked goods, fruit juices, syrups, olives and pickled sour vegetables.

10.5.3 Inorganic Antimicrobial

10.5.3.1 Sulfites and Sulfur Dioxide

Sulfur dioxide (SO_2) and its derivatives have been long used in foods as general food preservatives. They are added to food to inhibit non-enzymic browning, to inhibit enzyme catalyzed reactions, to inhibit and control microorganisms and to act as an antioxidant and a reducing agent. In general, SO_2 and its derivatives are metabolized to sulfate and excrete in the urine without any obvious pathologic results. However, because of somewhat recently recognized severe reactions to sulfur dioxide and its derivatives by some sensitive asthmatics, their use in foods is currently regulated and subject to rigorous labeling restrictions. Nonetheless, these preservatives are key roles in contemporary foods.

The commonly used forms in foods include sulfur dioxide gas and the sodium, potassium, or calcium salts of sulfite (SO^{2-}), bisulfite (HSO_3^-), or metabisulfde ($S_2O_5^{2-}$). The most frequently used sulfiting agents are the sodium and potassium metabisulfites since they show good stability toward autoxidation in the solid phase. However, gaseous sulfur dioxide is employed where leaching of solids causes problems or where the gas may also serve as an acid for the pH control.

Although the traditional names for the anions of these salts are still widely used (sulfites, bisulfltes, and metabisulfites), they have been designated by IUPAC as the sulfur oxoacids, sulfites (SO_3^{2-}), hydrogen sulfites (HSO_3^-), and disulfites ($S_2O_5^{2-}$), respectively. The oxoacids, H_2SO_3 and $H_2S_3O_5$, are designated as sulfurous and disulfurous acids, respectively.

Widely held views also have changed somewhat on the existence of sulfurous acid in aqueous solutions. Earlier, it was assumed that when sulfur dioxide was dissolved in water, it formed sulfurous acid, because the salts of simple oxoanions of sulfur (IV) (valence+4) are salts of this acid (H_2SO_3; sulfurous acid). However, evidence for the existence of free sulfurous acid has not been found, and it has been estimated that it accounts for less than 3% of nondissociated dissolved SO_2. Instead, solution of SO_2 yields only weak interactions with water, which results in a nondissociated complex that is particularly abundant below pH 2. This complex has been denoted $SO_2 \cdot H_2O$, and a distinction between this complex and sulfurous acid is not generally made.

10.5.3.2 Nitrite and Nitrate Salts

The potassium and sodium salts of nitrite and nitrate are often used in curing mixtures for meats to create and fix the color, to inhibit microorganisms, and to develop characteristic flavors. Nitrite rather than nitrate is obviously the functional constituent. Nitrites in meat form nitric oxide, which reacts with heme compounds to form nitrosomyoglobin, the pigment which is responsible for the pink color of cured meats. Sensory evaluations also suggest that nitrite contributes to cured meat flavor, apparently through an antioxidant role, but the details of this chemistry are still not clear yet. Furthermore, nitrites (150-200 ppm) inhibit *Clostridia* in canned-comminuted and cured meats. In this regard, nitrite shows relatively high antimicrobial activity at pH 5.0-5.5. The antimicrobial mechanism of nitrite is unknown, but it has been proposed that nitrite reacts with sulfhydryl groups to create compounds that are not metabolized by microorganisms under anaerobic conditions.

Nitrites have been shown to be involved in the formation of low, but possibly toxic levels of nitrosamthes in certain cured meals. Nitrite salts also occur naturally in many foods, including vegetables such as spinach. The accumulation of large amounts of nitrate in plant tissues grown on heavily fertilized soils is of concern, particularly in infant foods prepared from these tissues. The reduction of nitrate to nitrite in the intestine, with subsequent absorption, could lead to cyanosis due to methemoglobin formation. For these reasons, the use of nitrites and nitrates in foods has been doubted. The antimicrobial ability of nitrite provides some justification for its use in cured meats, especially when growth of *Clostridium botulinum* is possible.

10.5.4 Biological Antimicrobial

10.5.4.1 Nisin

The use of antibiotics in food preservation raises a problem since it might trigger development of more resistant microorganisms and thus create medical/therapeutic difficulties.

Nisin is a polypeptide composed of 34 amino acids, and is produced by some *Lactococcus lactis* strains. It can strongly inhibit Gram-positive microorganisms and all spores, but it is not used in human medicine. This heat-resistant peptide is employed as an additive for the sterilization of dairy products, such as cheeses or condensed or evaporated milk.

10.5.4.2 Natamycin

NATAMYCIN (I)

Natamycin or pimaricin is a polyene macrolide antimycotic that has been approved in the United States for use against molds on cured cheeses. This mold inhibitor is highly effective when applied to surfaces of foods exposed directly to air where mold is easy to proliferate. Natamycin is especially attractive for application on fermented foods, because it selectively inhibits molds while allowing normal growth and metabolism of ripening bacteria.

10.6 Sweeteners

10.6.1 Saccharin

Saccharin (3-oxo-2,3-dihydro-1,2-benziso-thiazole-1,1-dioxide) is usually available as the sodium salt and sometimes as the calcium salt as nonnutritive sweeteners.

SACCHARIN

The commonly accepted rule of thumb is that saccharin is about 300 times as sweet as sucrose in concentrations up to the equivalent of a 10% sucrose solution, but the range is from 200 to 700 times the sweetness of sucrose depending on the concentration and the food matrix. Saccharin exhibits a bitter, metallic aftertaste, especially to some individuals, and this becomes more evident with increasing concentration.

The safety of saccharin has been investigated for over 50 years, and it has been found to cause a low incidence of carcinogenesis in laboratory animals. However, many scientists argue that the animal data are not relevant to humans. In humans saccharin is absorbed, and then is rapidly excreted in the urine. Although current regulations in the United States prohibit the use of food additives that cause cancer in any experimental animals, a ban on saccharin in the United States has been stayed by the congressional legislation pending further researches, as proposed by the FDA in 1977. Saccharin has been assigned an ADI of 1.5 mg/kg of body weight. However, there are fears currently that this level may be exceeded by some sectors of the population, and the Ministry of Agriculture, Fisheries and Food (MAFF) in the UK have called for an investigation.

10.6.2 Aspartame

Aspartame or L-aspartyl-L-phenylalanine methyl ester (Fig. 10.6) is a caloric sweetener because it is a dipeptide that is completely digested after consumption. However, its intense sweetness (about 200 times as sweet as sucrose) allows functionality to be achieved at very low levels that provide insignificant calories. It is noted for a clean, sweet taste that is similar to that of sucrose. Aspartame was first approved in the United States in 1981, and now is approved for use in over 75 countries where it is used in over 1,700 products.

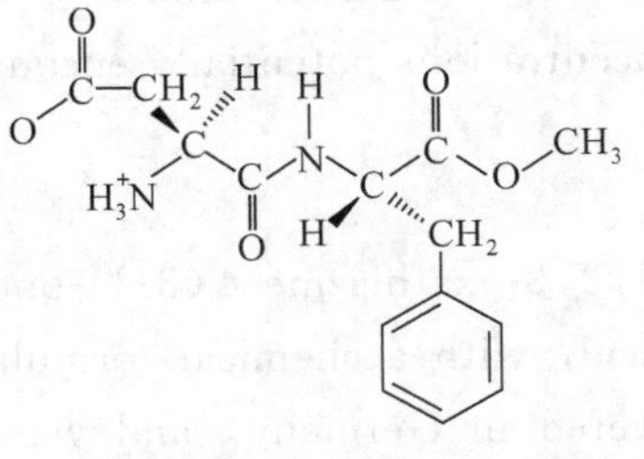

L-Aspartyl-L-Phenylalanine
Methyl Ester
(Aspartame)

Fig. 10.6 Stereochemical configuration of aspartame

Two disadvantages of aspartame are its instability under acid conditions and its rapid degradation when exposed to elevated temperatures. Under acid conditions, such as carbonated soft drinks, the rate of loss of sweetness is gradual and depends on temperature and pH. The peptide nature of aspartame makes it susceptible to hydrolysis, and this feature also permits other chemical interactions and microbial degradations. In addition to loss of sweetness resulting from hydrolysis of either the methyl ester on phenylalanine or the peptide bead between the two amino acids, aspartame readily undergoes an intramolecular condensation, especially at elevated temperatures, to yield the diketopiperazine (5-benzyl-3,6-dioxo-2-piperazine acetic acid) shown in Fig. 10.7. This reaction is especially favored at neutral and alkaline pH values, because nonprotonated amine groups on the molecule are more available for reaction under these conditions. Similarly, alkaline pH values promote carbonyl amino reactions, and aspartame has been shown to react readily with glucose under such conditions, loss of aspartame's sweetness during the storage is the principal concern.

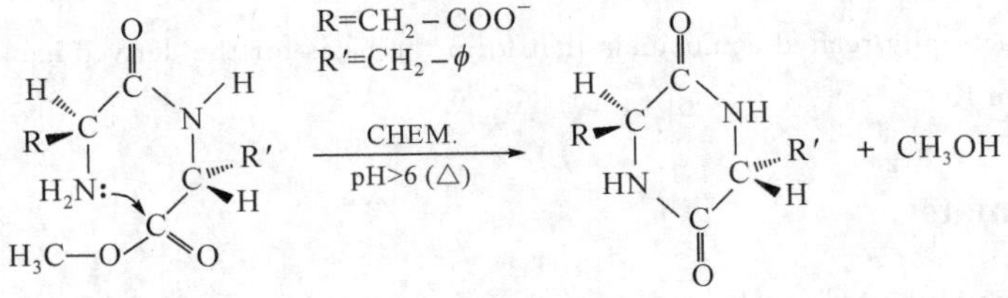

Aspartame → A Diketopiperazine + CH_3OH

$R = CH_2-COO^-$
$R' = CH_2-\phi$
CHEM.
pH>6 ($\triangle$)

Fig. 10.7 Intramolecular condensation of aspartame yielding a diketopiperazine degradation product

As a food additive, aspartame is also concerned about its potential safety, although it is composed of naturally-occurring amino acids and its daily intake is projected to be very small (0.8 g/person). Aspartame-sweetened products must be labeled prominently about their phenylalanine content to allow avoidance of consumption by phenylketonuric individuals who lack 4 monooxygenase that is involved in the metabolism of phenylalanine. However, consumption of aspartame by the normal population is not associated with adverse health effects. The extensive testing has similarly shown that the diketopiperazine poses no risk to humans at concentrations potentially encountered in foods.

10.6.3 Acesulfame K

Acesulfame K [6-methyl-1,2,3-oxathiazine 4(3H)-one-2,2-dioxide] is a potassium salt derived from acetoacetic acid, with a chemical formula of C_4H_4NOKS and a molar mass of 201.2. It was discovered in Germany, and was first approved for use as a nonnutritive sweetener in the U.S. in 1988. The complex chemical name of this substance led to the creation of the trademarked common name, Acesulfame K, which is based on its structural relationships to acetoacetic acid and sulfamic acid, and to its potassium salt nature (Fig. 10.8).

Acesulfame K is about 200 times as sweet as sucrose at a 3% concentration in solution, and it exhibits a sweetness quality between that of cyclamates and saccharin. Since acesulfame K possesses some metallic and bitter taste at higher concentrations, it is especially useful when mixed with other low calorie sweeteners, such as aspartame. Acesulfame K is exceptionally stable at elevated temperatures encountered in baking, and it is also stable in acidic products, such as carbonated soft drinks. Acesulfame K is not metabolized in the body, thus providing no calories, and is excreted by the kidneys unchanged. The extensive testing has shown no toxic effects in animals, and exceptional stability in food applications.

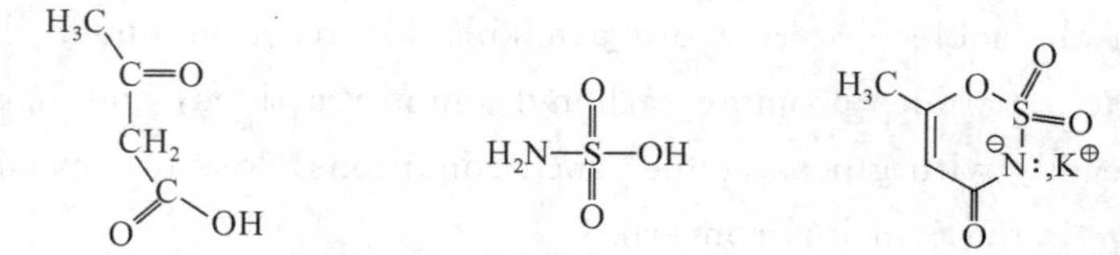

Aceto-Acetic Acid   Sulfemic Acid   Acesulfame K

Fig. 10.8 Structurally related compounds that form the basis for the derived name of the nonnutritive sweetener Acesulfame K

10.6.4 Cyclamate

Cyclamate (cyclohexyl sutfamate) was approved as a food additive in the U.S. in 1949, and before the substances were prohibited by the U.S. Food and Drug Administration in late 1969, the sodium and calcium salts and the acid form of cyclamic acid were widely used as sweeteners. Cyclamates are about 30 times sweeter than sucrose, taste much like sucrose without significant interfering taste sensations, and are heat stable. Cyclamate sweetness has slow onset, and persists for a period of time that is longer than that of sucrose.

Some early experimental evidence with rodents had suggested that cyclamate and its hydrolysis product, cyclohexylamine (Fig. 10.9), caused bladder cancer. However, the subsequent extensive testing has not substantiated the early reports, and petitions have been filed in the United States for reinstatement of cylcamate as an approved sweetener. Currently, cyclamate is permitted for use in low calorie foods in 40 countries. Still for various reasons, even though extensive data supporting the conclusion that neither cyclamate nor cyclohexylamine are carcinogen or genotoxic, the U. S. Food and Drag Administration has chosen not to reapprove cyclamates for use in foods.

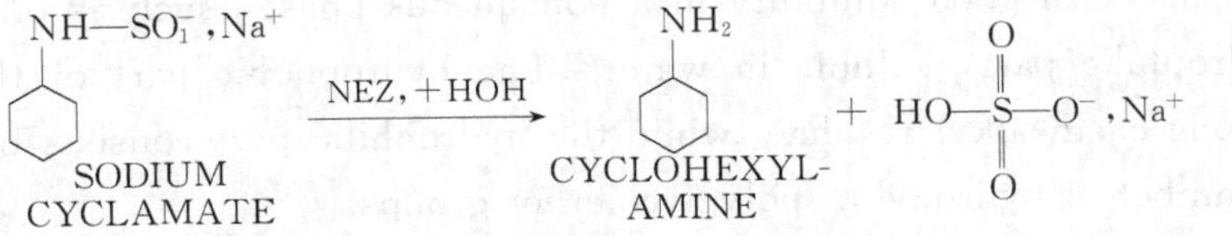

Fig. 10.9 Formation of cyclohexylamine by the hydrolysis of cyclamate

10.7 Emulsifier

Today, 150,000-200,000 tons of emulsifiers are produced worldwide. Of this amount, mono- and diacylglycerides and their derivatives account for the largest part, about 75%. Synthetic emulsifiers include a series of non-ionic compounds. Unlike the ionic compounds, the non-ionic emulsifiers are not in danger of decreasing in interfacial activity through the salt formation with food constituents. The utilization of emulsifiers is legislated and differently regulated in some countries. The synthetic emulsifiers described below are used worldwide.

Mono-and diacylglycerides, which are mostly used as mixtures, are produced as described. Other emulsifiers with special activities are obtained by the derivatization. As a result of the diverse reaction possibilities of the starting compounds, complex products are produced in this process.

Sugar esters are obtained, among other methods, by transesterification of fatty acid methyl esters (14:0, 16:0, 18:0 and/or 18:1, double bond position 9) with sucrose and lactose. The resultant mono-and diesters are odorless and tasteless. Depending on their structure, they cover an HLB range of 7-13, and are used in stabilization of o/w emulsions, or in stabilization of some instant dehydrated and powdered foods.

Esters of sorbitan with fatty acids (Span's) serve the stabilization of w/o emulsions.

Sorbitan tristearate is often used in the production of chocolate to delay the fat bloom formation.

Polyoxyethylene groups are introduced into the molecules to increase the hydrophilic property of sorbitan esters. Polyoxyethylene sorbitan monoesters (Tween's) are used to stabilize o/w emulsions.

10.7.1 Emulsifier Action

10.7.1.1 Structure and Activity

Emulsions are made and stabilized with the aid of a suitable tenside, usually called an emulsifier. Its activity depends on the molecular structure. There is a lipophilic or hydrophobic part with good solubility in a nonaqueous phase, such as oil or fat, and a polar or hydrophilic part, soluble in water. The hydrophobic part of the molecule is generally a long-chain alkyl residue, while the hydrophilic part consists of a dissociable group or a number of hydroxyl or polyglycolether groups.

In an immiscible system such as oil/water, the emulsifier is located on the interface, where it decreases the interfacial tension. Thus, even in a very low concentration, it facilitates a fine distribution of one phase within the other. The emulsifier also prevents droplets, once formed, from aggregating and coalescing, i. e., merging into a single, large drop (Fig. 10.10). Ionic tensides stabilize o/w emulsions in the following way (Fig. 10.11a): at the interface, their alkyl residues are solubilized in oil droplets, while the charged end groups project into the aqueous phase. The involvement of counter ions forms an electrostatic double layer, which prevents the oil droplet aggregation.

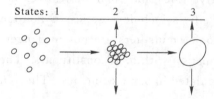

Fig. 10.10 Changes in an emulsion. 1. The droplets are dispersed in a continuous phase. 2. The droplets form aggregates. An increase in particle diameter results in acceleration of their flotation or sedimentation. 3. Coalescence: the aggregated droplets merge into larger droplets. Finally, two continuous phases are formed; the emulsion is destroyed

Nonionic, neutral tensides are oriented on the droplet's surface with the polar end of the tenside projecting into the aqueous phase. The coalescence of the droplets of an o/w emulsion is prevented by an anchored "hydrate shell" built around the polar groups.

The coalescence of water droplets in a w/o emulsion first requires that water molecules break through the double-layered hydrophobic region of emulsifier molecules (Fig. 10.11(b)) This escape is only possible when sufficient energy is supplied to rupture the emulsifier's hydrophobic interaction.

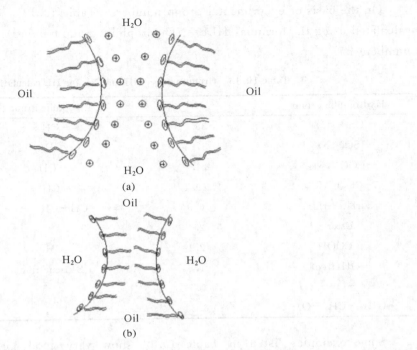

Fig. 10.11 Stabilization of an emulsion. (a) Activity of an ionic emulsifier in an o/w emulsion. (b) Activity of a nonpolar emulsifier in w/o emulsion.

The stability of an emulsion is increased when additives are added which curtail droplet mobility. This is the basis of the stabilization effect of hydrocolloids on o/w emulsions since they increase the viscosity of the outer, aqueous phase.

A rise in temperature negatively affects the emulsion stability, and can be applied whenever an emulsion has to be destroyed. Elevated temperature are used along with shaking, agitation or pressure (mechanical destruction of interfacial films as, for example, in butter manufacturing). Other ways of decreasing the stability of an emulsion are addition of ions which collapse the electrostatic double layer, or hydrolysis to destroy the emulsifier.

10.7.1.2 HLB-Value

A tenside with a relatively strong lipophilic group and weak hydrophilic group is mainly soluble in oil and preferentially stabilizes a w/o emulsion, and vice versa. This fact leads to the development of a standard with which the relative strength or 'activity' of the hydrophilic and lipophilic groups of emulsifiers can be enhanced. It is called the HLB value (hydrophilic-lipophilic balance). It can be determined, e.g., from dielectric constants or from the chromatographic behavior of the surface-active substance. The HLB value of the fatty acid esters of polyhydroxy alcohols can also be calculated as follows (SV=saponification number of the emulsifier, AV=acid value of the separated acid): HLB=20(1-SV/AV).

FOOD CHEMISTRY

On the basis of experimental group numbers (Table 10.13), the HLB value can be calculated using the formula: HLB=Σ(hydrophilic group number)−Σ(hydrophibic group number)+7.

Table 10.13 Group number NH and NL for HL calculation

Hydrophilic group	NH	Lipophilic group	NL
$-OSO_3^-$, Na^+	38.7	$-CH-$	0.475
$-SO_3^-$, Na^+	37.4	$-CH_2-$	0.475
$-COO^-$, Na^+	21.1	$-CH_3$	0.475
$-COO^-$, K^+	19.1	$=CH-$	0.475
Sorbitan ring	6.8	$-CH-CH_3-O-$	0.15
Ester	2.4		
$-COOH$	2.1	CH_3	
$-OH$(free)	1.9	Benzene ring	1.662
$-O-$	1.3		
$-(CH_2-CH_2-O)-$	0.33		

Some examples listed in Table 10.14 show very good correspondence between calculated and experimentally found HLB values.

Table 10.14 Hydrophilic lipophilic balance (HLB) values of some surfactants

Compound	HLB-value	
	Found	Calculated
Oleic acid	1.0	
Sorbitol tristearate	2.1	2.1
Stearyl monoglyceride	3.4	3.8
Sorbitol monostearate	4.7	4.7
Sorbitol monolaurate	8.6	
Gelatin	9.8	
Polyoxyethylene sorbitol tristearate	10.5	10
Methylcellulose	10.5	
Polyoxyethylene sorbitol monostearate	14.9	
Polyoxyethylene sorbitol monooleate	15.0	15
Sodium oleate	18.0	
Potassium oleate	20.0	

The HLB values suggested the first industrial applications (Table 10.15). For a detailed characterization, however, comprehensive knowledge of possible interactions of the emulsifier with the many components of a food emulsion is still lacking. Hence, emulsifiers are mainly used in accordance with empirical considerations.

Table 10.15 HLB-values related to their industrial application

HLB-range	Application
3-6	w/o-Emulsifiers
7-9	Humectants
8-18	ow-Emulsifiers
15-18	Turbidity stabilization

It has been observed with neutral emulsifiers that the degree of hydration of the polar groups decreases with a rise in temperature and the influence of the lipophilic groups increases. The phase inversion occurs o/w → w/o. The temperature at which inversion occurs is called the phase conversion temperature.

10.7.2 Synthetic Emulsifiers

10.7.2.1 Mono-, Diacylglycerides and Derivatives

Mono- and diacylglycerides, which are mostly used as mixtures (Table 10.16), are produced. Other emulsifiers with special activities are obtained by the derivatization. As a result of the diverse reaction possibilities of the starting compounds, complex products are obtained in this process. An example is represented by the diacetyltartaric acid ester of monoglycerides (DATEM). At concentrations of ca. 0.3% (based on the amount of flour), this ester increases the volume of wheat biscuits. For the production of this emusifier, acetic anhydride and tartaric acid are heated, diacetyltartaric acid anhydride being formed on the removal of acetic acid by the distillation.

Table 10.16 Emulsifiers from mono-and diacylglyceride mixtures

Mono-and diglycerides esterified with	Name	EU-number	Production by conversion of mixtures of mono-and diacylglycerides with
Acetic acid (acetylated mono-and diglycerides)	Acetem	E472a	Acetic anhydride
Lactic acid	Lactem	E472b	Lactic acid
Citric acid	Citrem	E472c	Citric acid
Monoacetyl-and diacetyltartaric acid	Datem	E472e	Tartaric acid and acetic anhydride

10.7.2.2 Sugar Esters

Suger esters obtained by the transesterification of fatty acid methyl esters (14:0, 16:0, 18:0 and/or 18:1, double bond position 9) with sucrose and lactose. The resultant mono-and diesters are odorless tasteless. Depending on their structures they cover an HLB range of 7-13, and are use in stabilization of o/w emulsions, or in the stabilization of some instant dehydrated and powdered foods.

10.7.2.3 Sorbitan Fatty Acid Esters

Esters of sorbitan with fatty acids (Span's) serve the stabilization of w/o emulsions.

Sorbitan tristearate is used in the production of chocolate to delay the fat bloom formation.

10.7.2.4 Polyoxyethylene Sorbitan Esters

Polyoxyethylene groups (Tween's) are introduced into the molecules to increase the hydrophilic property of sorbitan esters:

Polyoxyethylene sorbitan monoesters (examples in Table 10.14) are used to stabilize o/w emulsions.

10.7.2.5 Polyglycerol-Polyricinoleate (PGPR)

In the production of the emulsifier PGPR, oligomeric glycerol is made by attachment of 2,3-epoxy-1-propanol (glycid) to glycerol and at the same time ricinoleic acids are esterified with each other under controlled heating conditions. In a third step, the oligomeric glycerol is esterified with poly-esterricinoleic acid.

The emulsifier has a very complicated composition apart from different types of esters, oligomeric glycerol and free ricinoleic acid are present. Together with lecithin, PGPR is used in the production of chocolate. It completely eliminates the flow point of a molten chocolate.

R=H, ricinoleic acid or polyricinoleic acid

10.7.2.6 Stearyl-2-Lactylate

In the presence of sodium or calcium hydroxide, esterification of stearic acid with lactic acid gives a mixture of stearyl lactylates (Na or Ca salt), the main component being stearyl-2-lactylate:

$$CH_3-(CH_2)_{16}-CO-O-\underset{\underset{CH_3}{|}}{CH}-CO-O-\underset{\underset{CH_3}{|}}{CH}-COOH^{\ominus}\ Na^{\oplus}$$

The free acid acts as a w/o emulsifier and the salts as o/w emulsifiers. The HLB value of the sodium salt is 8-9, and that of the calcium salt, 6-7. The sodium salt is used to stabilize an o/w emulsion which is subjected to repeated cycles of freezing and thawing.

10.8 Antioxidants

Since lipids are widely distributed in food and since their oxidation yields degradation products of great aroma impact, their degradation is an important cause of food deteriorations by the generation of undesirable aroma. The lipid oxidation can be retarded by the oxygen removal or by using antioxidants as food additives. The latter are mostly phenolic compounds, which provide the best results often as a mixture and in combination with a chelating agent. The most important antioxidants, natural or synthetic, are tocopherols, ascorbic acid esters, gallic acid esters, tert-butylhydroxyanisole (BHA) and di-tert-butylhydroxytoluene (BHT).

Glossary

acetic	乙酸的
acidulant	酸化剂
ADI	每日允许摄入量
additive	添加剂
alkaline	碱性的
antibiotic	抗生素
antimycotic	抗真菌的
antioxidant	抗氧化剂
aspartame	阿斯巴甜
benzoic	安息香的

bitter	苦的
buffer	缓冲
citrate	柠檬酸盐
cyclamate	甜蜜素
disulfide	二硫化物
emulsifier	乳化剂
equilibrium	平衡
HLB	亲水亲油平衡值
hydrocolloid	水状胶质
hydronium	水合氢离子
hydrophobic	疏水的
interface	界面
labile	不稳定的
lecithin	卵磷脂
monoglyceride	单甘油酯
natamycin	纳他霉素
neutralize	中和
nisin	乳酸链球菌肽
nitrite	亚硝酸盐
paraben	尼泊金（苯甲酸酯类）
phosphoric	含磷的
propionic	丙酸的
pudding	布丁
racemization	外消旋作用
rennet	凝乳酶
saccharin	糖精
sorbic	山梨酸的
sulfhydryl	含巯基的
sulfite	亚硫酸盐
sweetener	甜味剂
tartrate	酒石酸盐
tenside	表面活性剂
tocopherol	生育酚
undissociated	未解离的

References

Belitz HD, Chen W, Jugel H, Stempfl W, Treleano R, Wieser H. *Structural Requirements for Sweet and Bitter Taste*. In: Schreier, P (Ed.), Flavour. Berlin: Walter de Gruyter, 1981: 741.

Birch GG, Brennan JG, Parker KJ. *Sensory Properties of Foods*. London: Applied Science Pubs, 1977.

Blair J. S. *Color stabilization of Green Vegetables*. U. S. Pat. 1940: 2,186, 003.

Blow DM, Birktoft JJ, Hartley BS. Chymotrypsin. *Nature*, 1969, 221: 337-340.

Branen AL, Davidson PM, Salminen S. *Food Additives*. New York: Marcel Dekker, 1990.

Chahine MH, Deman JM. Autoxidation of corn oil under influence of fluorescent light. *Canadian Institure of Food Science and Technology Journal*, 1971, 4(1): 24.

Creighton TE. Proteins: *Structures and Molecular Properties*. New York: WH Freeman, 1983.

Fox JB, Jr. The Chemistry of Meat Pigments. *Journal of Agricultural and Food Chemistry*, 1996, 14: 207-210.

Furia TE. *Handbook of Food Additives*. Second edition. Cleveland, Ohio: CRC Press, 1972.

Geis I. *Biochemistry*(2nd ed.)(F. B. Armstrong). New York: Oxford University Press, 1983: 108.

Gould GW. *Mechanisms of Action of Food Preservation Procedures*. London: Elsevier Applied Science, 1989.

Griffin WC, Lynch MJ. *Surface Active Agents*. In Furia TE (Ed.), Handbook of Food Additives, second endorse. Cleveland, Ohio: CRC Press, 1972: 397.

Ha, J. K. , and R. C. Lindsay. Volatile alkylphenols and thiophend in species-related characterizing flavors of red meats. *J Food Sci*, 1991, 56: 1997-1202.

Hans Lineweaver, Dean Burk. The determination of enzyme dissociation consants. *Journal of the American cherical Society*, 1934,56(3): 658-666.

Herbert RA. *Microbial Growth at Low Temperatures*. In Gould GW (Ed.), Mechanism of action of food preservation procedures. London: Elsevier Applied Science, 1989: 71.

Hudson BJF. *Developments in Food Proteins*-1. London: Applied Science Publication, 1982.

James S, Blalar, Maywood. *Color Stabilization of Green Vegetables*. United States Patent Office, 1940:262,305.

Lawrie RA. *Meat Science*(4th ed.). New York: Pergamon Press, 1985.

Leistner L. *Principles and Applications of Hurdle Technolgy*. In Gould GW(Ed.), New Methods of Food Preservation, Springer, 1995.

Lucca PA, Tepper BJ. Fat replacers and the functionality of fat in foods. *Trends in Food Science & Technology*, 1994, 5, 12-19.

Morrissay PA, Mulvihill DM, O'Neill EM. *Functional Properties of Muscle Proteins*. In: Hudson BJF (Ed.), Development in Food Proteins, vol. 5. London: Elsevier Applied Science, 1987: 195.

Na, J. K, and R. C. Lindsay. Volatile alkylphenols and thiophenol in species-related characterizing flavors of red. *J Food Sci*, 1991, 56: 1197-1202.

Owen R. Fennema. *Food Chemistry*. New York-Basec: Marcel Dekker Inc, 1997.

Peng CY, Markakis P. Effect of prenolase on anthocyanins. *Nature*(Lond.), 1963, 199:597-598.

Perutz MF. *Proteins and Nucleic Acids*. Amsterdam: Elsevier Publication Corporation, 1962.

Powrie WD, Tung MA. *Food Dispersions*. In: Fennema OR (Ed.), Principles of food science, part I. New York: Marcel Dekker, Inc, 1976: 539.

Sattar A, Deman JM, and Alexander JC. Stability of edible oils and fats to fluorescent light irradiation. *Journal of the American oil chemists society*, 1976, 53(7): 473-477.

Schulz GE, Schirmer RH. *Principle of Protein Structure*. Berlin-Heidelberg: Springer-Verlag, 1979.

Tschesche H. *Modern Methods in Protein Chemistry*, vol. 2. Berlin: Wather de Gruyter, 1985.

Walton AG. *Polypeptides and Protein Structure*. New York-Oxford: Elsevier North Holland, 1981.

Witaker. *Principces of Enzymology for the Food Science*. New York-Basel: Marcel Dekker, Inc, 1994.

Wong, E., L. N. Nixon, and C. B. Johnson. Volatile mdium chain fatty acids and mutton flavor. *J Agric Food Chem*. 1975, 23: 495-498.